# MATHEMATICAL METHODS FOR ENGINEERING SCIENCE

BY

DR. Peta Trigger  Ph.D, Ed.D
Northampton Academy of Post Doctoral Studies

K B P
2 Emms Hill Barns
Hamsterley
County Durham
First published 2013

ISBN 9781494815660

PRINTED BY CREATESPACE
https://www.createspace.com

## FOREWORD

This book develops 3 mathematical methods of solving problems in engineering science which involve the inversion of matrices with complex elements; the solution of higher order simultaneous equations; and the solution of higher order polynomial equations. The author has used these methods as the rationale for writing computer programs which can be used to handle situations where the higher order equations involved would result in long, tedious and error-prone calculations, particularly in electrical engineering where component values tend not be conveniently small and simple integer quantities.

The mathematical background of each method is developed from first principles. Examples of the types of problems which these may be used to solve are also described.

The solutions obtained may be checked for veracity as follows:

In the matrix inversion method, the inverted matrix found is multiplied by the original matrix to determine how close the product comes to the identity matrix.

In the simultaneous equations-solving method, the solution found is substituted into the set of simultaneous equations, and the r.h.s.'s are calculated to determine how close they come to the constants in each equation.

In the polynomial equation-solving method, the solutions found are substituted into the polynomial equation to determine how close the result comes to zero.

# TABLE OF CONTENTS

LIST OF FIGURES 7.

**1. INVERTING MATRICES WITH COMPLEX ELEMENTS** 9.

INTRODUCTION 9.

TRANSMISSION MATRICES: DEFINITION 11.

SOLVING A.C. CIRCUITS FOR CURRENTS AND VOLTAGES USING TRANSMISSION MATRICES 11.

METHODS OF INVERTING MATRICES WITH REAL ELEMENTS ONLY 23.

DETERMINANTS: DEFINITION 25.

CONDENSATION METHODS OF EVALUATING DETERMINANTS 28.

ADJOINT MATRICES 36.

AN EXPRESSION FOR THE INVERTED MATRIX IN TERMS OF |**A**| AND ADJ **A** **38**.

METHODS OF INVERTING MATRICES WITH COMPLEX ELEMENTS 41.

DIVISION OF COMPLEX NUMBERS 44.

APPLICATION OF THE METHOD TO A PRACTICAL PROBLEM IN ENGINEERING SCIENCE 48.

**2. SOLVING HIGHER ORDER SIMULTANEOUS EQUATIONS 57.**

INTRODUCTION 57.

SOLVING D.C. CIRCUITS FOR CURRENTS AND VOLTAGES 57.

SOLUTION OF SIMULTANEOUS EQUATIONS BY GAUSS-JORDAN ELIMINATION 60.

SOLVING SIMULTANEOUS EQUATIONS
USING ADJOINT MATRICES AND
DETERMINANTS                63.

APPLICATION OF THE METHOD TO A
PRACTICAL PROBLEM IN ENGINEERING
SCIENCE                     70.

## 3. SOLVING POLYNOMIAL EQUATIONS  77.

INTRODUCTION                77.

THE TRANSFER CHARACTERISTIC OF A
D.C. AMPLIFIER              77.

SOLVING POLYNOMIAL EQUATIONS  78.

A METHOD OF SOLVING POLYNOMIAL
EQUATIONS USING THE NEWTON-RAPHSON
APPROACH                    79.

FINDING SUBSEQUENT SOLUTIONS  80.

FINDING COMPLEX SOLUTIONS          83.

APPLICATION OF THE METHOD TO A PRACTICAL PROBLEM IN ENGINEERING SCIENCE          84.

CONCLUSION          90.

**APPENDICES**          **93.**

APPENDIX 1: FINDING THE FORM OF TRANSMISSION MATRIX FOR CIRCUIT STAGES CONTAINING AN IMPEDANCE, A RESISTANCE, AN ADMITTANCE AND A CONDUCTANCE          93.

APPENDIX 2: MATRIX MULTIPLICATION          98.

APPENDIX 3: DETERMINANT PROPERTIES          102.

## LIST OF FIGURES

FIG 1: AN A.C. CIRCUIT          9.

FIG 2: SHOWING THE POLAR REPRESENTATION OF THE COMPLEX NUMBER a + ib as r.(cosθ + isinθ)
         46.

Fig 3: A 'π' NETWORK CONSISTING ON AN IMPEDANCE AND TWO ADMITTANCES
         48.

Fig. 4: A CIRCUIT CONSISTING OF RESISTANCES AND A VOLTAGE SOURCE
         57.

FIG. 5 : A CIRCUIT CONTAINING RESISTANCES AND A VOLTAGE SOURCE
         70.

FIG 6 : THE TRANSFER CHARACTERISTIC OF A D.C. AMPLIFIER
         77.

FIG 7: CIRCUIT DIAGRAM OF A PUSH-PULL AMPLIFIER          85.

# 1. INVERTING MATRICES WITH COMPLEX ELEMENTS

INTRODUCTION

The aim of this introduction is to compare solutions giving the voltages and currents in a circuit using voltage and current laws, with a method involving multiplication of transmission matrices.

Consider the A.C. circuit in Fig. 1(a) below

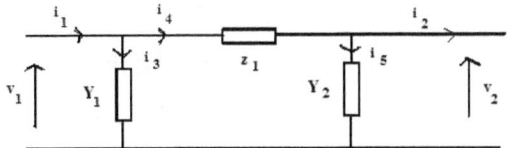

FIG.1 (a) COMPLETE CIRCUIT

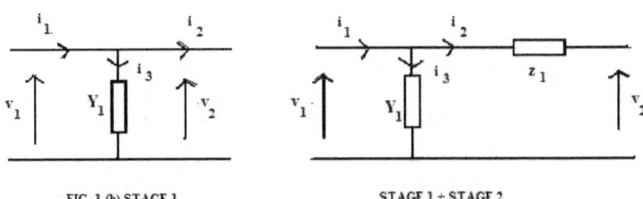

FIG. 1 (b) STAGE 1          STAGE 1 + STAGE 2

FIG 1: AN A.C. CIRCUIT

A method involving multiplication of transmission matrices will be shown to be advantageous when compared the use of voltage and current laws to calculate the voltages and currents in the circuit.

Although the matrices involved contain only two rows and two columns, the aim will be to develop a general method which can be used with more complex problems involving 3 x 3 and higher order matrices. This will then be used to solve the simpler 2 x 2 problem by way of example.


DEFINITION OF A MATRIX

A matrix is a two-dimensional array of rows and columns of numbers or variables called 'elements'.
'Square' matrices, the type of interest here, have equal numbers of rows and columns. A column matrix has a single column of elements.


Matrices have their own particular set of rules (which, when they are used here will not be proved but simply quoted), for example for addition, multiplication and division. However, the method used to multiply two matrices is important in solving the problems of interest here and so matrix multplication is described in Appendix 2.

## TRANSMISSION MATRICES

A transmission matrix, of special interest here, is a 2 x 2 (i.e. 2 rows and 2 columns) matrix with elements representing resistances, conductances, impedances or admittances.

Fig. 1 on p. 9 shows a general example of a circuit with components whose values may be represented by the elements in a 2 x 2 transmission matrix.

## THE FIRST STAGE OF THE CIRCUIT

Looking at the first stage of the circuit in Fig. 1 (b), by the Kirchhoff Current Law (that the algebraic sum of currents into and out of a junction is zero), we have:

$$i_1 - i_2 - i_3 = 0,$$

taking the convention that currents into a junction are positive, those out of a junction are negative.

But $i_3 = -v_2.Y_1$ (Y denotes an admittance).

And so $i_1 = v_2.Y_1 + i_2$ \quad (1)

Now the component in Stage 1 can be represented by the transmission matrix (see Appendix 1):

$$\begin{bmatrix} 1 & 0 \\ Y_1 & 1 \end{bmatrix}$$

and the output current and voltage is represented in matrix form by:

$$\begin{bmatrix} v_2 \\ i_2 \end{bmatrix}$$

Using the rule for multiplying matrices (to obtain the 'entry' or element $x_{ij}$, multiply each element in row i in the first matrix by the corresponding element in column j of the second matrix);

$$\begin{bmatrix} 1 & 0 \\ Y_1 & 1 \end{bmatrix} \begin{bmatrix} v_2 \\ i_2 \end{bmatrix} = v_2 . Y_1 + i_2$$

21

which is the same as equation (1).

'21' is the element in the second row and first

column of the product of the two matrices in the above equation.

$v_2$ is simply equal to $v_1$ which is equal to:

$$\begin{bmatrix} 1 & 0 \\ Y_1 & 1 \end{bmatrix} \begin{bmatrix} v_2 \\ i_2 \end{bmatrix} = v_2$$

11

## THE FIRST + SECOND STAGES OF THE CIRCUIT

Looking at the first and second stages, we have:

$$i_1 = i_2 + v_1 Y_1 \quad (2)$$

So $v_1 = i_2 . z_1 + v_2 \quad (3)$

and therefore $i_2 = \dfrac{v_1 - v_2}{z_1}$

Substituting for $v_1$ in (2), this gives:

$$i_1 = i_2 + [i_2 . z_1 + v_2] . Y_1$$

and so $i_1 = i_2 + v_2.Y_1 + Y_1.z_1.i_2$ (4)

Now $v_2 = v_1 - i_2.z_1$

Rearranging,
$v_1 = v_2 + i_2.z_1$

But

$$\begin{bmatrix} v_1 \\ i_1 \end{bmatrix} = \begin{bmatrix} 1 & 0 \\ Y_1 & 1 \end{bmatrix} \begin{bmatrix} 1 & z_1 \\ 0 & 1 \end{bmatrix} \begin{bmatrix} v_2 \\ i_2 \end{bmatrix}$$

where the second matrix is the transmission matrix for stage 2.

Multiplying out the first two matrices gives:

$$i_1 = \begin{bmatrix} 1 & z_1 \\ Y_1 & Y_1.z_1 + 1 \end{bmatrix} \begin{bmatrix} v_2 \\ i_2 \end{bmatrix}$$

and therefore

$$i_1 = \begin{bmatrix} 1 & z_1 \\ Y_1 & Y_1.z_1 + 1 \end{bmatrix} \begin{bmatrix} v_2 \\ i_2 \end{bmatrix}_{11}$$

$$= v_2.Y_1 + (Y_1.z_1 + 1).i_2$$
$$= v_2.Y_1 + Y_1.z_1.i_2 + i_2$$

which is equation (4) above.

Similarly,

$$v_1 = \begin{bmatrix} 1 & z_1 \\ Y_1 & Y_1.z_1 + 1 \end{bmatrix} \begin{bmatrix} v_2 \\ i_2 \end{bmatrix}_{21}$$

$$= v_1 = i_2.z_1 + v_2$$

which is equation (3).

## ALL THREE STAGES OF THE CIRCUIT

Finally, for the complete circuit, using Kirchhoff's current law;

$i_1 - i_4 - i_3 = 0$ (5)

$i_4 - i_2 - i_5 = 0$ (6).

Also

$i_3 = v_1 Y_1,$

$i_4 = \dfrac{v_1 - v_2}{z_1}$

and

$i_5 = v_2 . Y_2 .$

Substituting these last equations into equations (5) and (6):

$i_1 - \dfrac{v_1 - v_2}{z_1} - v_1 Y_1 = 0$ (7)

$\dfrac{v_1 - v_2}{z_1} - i_2 - v_2 . Y_2 = 0$ (8)

So, from (8)

$\dfrac{v_1 - v_2}{z_1} = i_2 - v_2 . Y_2$

and so

$$v_1 = v_2 + z_1 \cdot i_2 + v_2 \cdot z_1 \cdot Y_2 \qquad (9)$$

which is equal to

$$\begin{bmatrix} 1 & 0 \\ Y_2 & 1 \end{bmatrix} \begin{bmatrix} 1 & z_1 \\ 0 & 1 \end{bmatrix} \begin{bmatrix} 1 & 0 \\ Y_2 & 1 \end{bmatrix} \begin{bmatrix} v_2 \\ i_2 \end{bmatrix}_{11}$$

where

$$\begin{bmatrix} 1 & 0 \\ Y_2 & 1 \end{bmatrix}$$

is the transmission matrix for the 3rd stage.

Since

$$\begin{bmatrix} 1 & 0 \\ Y_1 & 1 \end{bmatrix} \begin{bmatrix} 1 & z_1 \\ 0 & 1 \end{bmatrix} = \begin{bmatrix} 1 & z_1 \\ Y_1 & Y_1 \cdot z_1 - 1 \end{bmatrix}$$

and

$$\begin{bmatrix} 1 & z_1 \\ Y_1 & Y_1 \cdot z_1 + 1 \end{bmatrix} \begin{bmatrix} 1 & 0 \\ Y_2 & 1 \end{bmatrix} = \begin{bmatrix} 1 + z_1 \cdot i_2 & z_1 \\ Y_1 - Y_1 \cdot z_1 \cdot Y_2 + Y_2 & Y_1 \cdot z_1 + 1 \end{bmatrix}$$

$$\begin{bmatrix} 1 + z_1 \cdot i_2 & z_1 \\ Y_1 + Y_1 \cdot z_1 \cdot Y_2 + Y_2 & Y_1 \cdot z_1 + 1 \end{bmatrix} \begin{bmatrix} v_2 \\ i_2 \end{bmatrix} = v_2 [1 + z_1 \cdot Y_2] + z_1 \cdot i_2$$

$$= v_2 + z_1 . i_2 + v_2 . z_1 . Y_2 \text{, as required.}$$

Substituting (8) and (9) in (7):
$$i_1 - [i_2 + v_2 . Y_2] - [v_2 + z_1 . i_2 + v_2 . z_1 . Y_2] Y_1 = 0$$

So
$$i_1 - v_2 . Y_1 - z_1 . i_2 . Y_1 - v_2 . z_1 . Y_2 . Y_1 - i_2 - v_2 . Y_2 = 0$$

From which
$$i_1 = v_2 . Y_1 + v_2 . Y_2 + v_2 . z_1 . Y_2 . Y_1 + i_2 + z_1 . i_2 . Y_1$$

which is equal to

$$\begin{bmatrix} 1 & 0 \\ Y_1 & 1 \end{bmatrix} \begin{bmatrix} 1 & z_1 \\ 0 & 1 \end{bmatrix} \begin{bmatrix} 1 & 0 \\ Y_2 & 1 \end{bmatrix} \begin{bmatrix} v_2 \\ i_2 \end{bmatrix}$$

Hence, for the complete circuit:

$$\begin{bmatrix} v_1 \\ i_1 \end{bmatrix} = \begin{bmatrix} 1 & 0 \\ Y_1 & 1 \end{bmatrix} \begin{bmatrix} 1 & z_1 \\ 0 & 1 \end{bmatrix} \begin{bmatrix} 1 & 0 \\ Y_2 & 1 \end{bmatrix} \begin{bmatrix} v_2 \\ i_2 \end{bmatrix}$$
(10)

which could be written down immediately with much reduced effort compared with the corresponding derivation using the current and voltage laws.

## FINDING THE OUTPUT VOLTAGE AND CURRENT FROM THE INPUT VOLTAGE AND CURRENT

If the first three matrices in (10) are multiplied out, we obtain an equation of the form

$$\begin{bmatrix} v_1 \\ i_1 \end{bmatrix} = \begin{bmatrix} a & b \\ c & d \end{bmatrix} \begin{bmatrix} v_2 \\ i_2 \end{bmatrix}.$$

In an operation corresponding to ordinary arithmetic,

$$\begin{bmatrix} v_1 \\ i_1 \end{bmatrix} \begin{bmatrix} a & b \\ c & d \end{bmatrix}^{-1} = \begin{bmatrix} v_2 \\ i_2 \end{bmatrix}$$

where

$$\begin{bmatrix} a & b \\ c & d \end{bmatrix}^{-1}$$

is the inverse of

$$\begin{bmatrix} a & b \\ c & d \end{bmatrix}$$

which, when values are assigned to the

admittances and impedances, will contain complex elements.

For example, $Y_1$ might be $3 + 1j$; $z_1$ $2 - 4j$; $Y_2$ $5 - 3j$; $v_1$ $-6 - 3.8j$ and $i_1$ $-14.4 - 19.1j$, with appropriate units. What are the values of $v_2$ and $i_2$ ?

The rule for addition of matrices allows us to split a complex matrix into a matrix with real coefficients and a matrix with imaginary coefficients:

Thus

$$\begin{bmatrix} a+ib \\ e+if \end{bmatrix} \begin{bmatrix} c+id \\ g+ih \end{bmatrix} = \begin{bmatrix} a & c \\ e & g \end{bmatrix} + \begin{bmatrix} ib & id \\ if & ih \end{bmatrix}$$

So $\begin{bmatrix} a+ib \\ e+if \end{bmatrix} \begin{bmatrix} c-id \\ g-ih \end{bmatrix} = \begin{bmatrix} a & c \\ e & g \end{bmatrix} + i\begin{bmatrix} b & d \\ f & h \end{bmatrix}$

Hence, the first step is to consider methods of inverting matrices with real elements only, from which the chosen method can be extended to encompass matrices with the addition of imaginary components to these real entries.

## METHODS OF INVERTING MATRICES WITH REAL ELEMENTS ONLY

A popular method of matrix inversion is the Gaussian method of row reduction. Its application to inverting a 3x3 matrix will be considered for simplicity, but the method is applicable to higher order matrices with corresponding jumps in the amount of labour involved.

Consider the matrix

$$A = \begin{bmatrix} 1 & 1 & 2 \\ -1 & 0 & -4 \\ 3 & 2 & 10 \end{bmatrix}$$

First, an 'augmented' matrix, **I** (an identity matrix containing 1's in the leading diagonal and 0's elsewhere), is juxtaposed with **A** and partitioned from it.

The rows are next labelled $r_1$, $r_2$, and $r_3$.

The aim of the method is to take each row in turn and add or subtract multiples of it to produce a zero element in the new row:

$$
\begin{array}{l}
r_1 \\
r_2 \\
r_3
\end{array}
\quad
\left( \begin{array}{ccc|ccc}
1 & 1 & 2 & 1 & 0 & 0 \\
-1 & 0 & -4 & 0 & 1 & 0 \\
3 & 2 & 10 & 0 & 0 & 1
\end{array} \right)
$$

$$
\begin{array}{l}
\\
r_2 \\
r_3 \to r_3 - 3r_1
\end{array}
\quad
\left( \begin{array}{ccc|ccc}
1 & 1 & 2 & 1 & 0 & 0 \\
0 & 1 & -2 & 1 & 1 & 0 \\
0 & -1 & 4 & -3 & 0 & 1
\end{array} \right)
$$

$$
\begin{array}{l}
r_1 \to r_1 - r_2 \\
\\
r_3 \to r_3 + r_2
\end{array}
\quad
\left( \begin{array}{ccc|ccc}
1 & 0 & 4 & 0 & -1 & 0 \\
0 & 1 & -2 & 1 & 1 & 0 \\
0 & 0 & 2 & -2 & 1 & 1
\end{array} \right)
$$

$$
\begin{array}{l}
\\
\\
r_3 \to 1/2\, r_3
\end{array}
\quad
\left( \begin{array}{ccc|ccc}
1 & 0 & 4 & 0 & -1 & 0 \\
0 & 1 & -2 & 1 & 1 & 0 \\
0 & 0 & 1 & -1 & \tfrac{1}{2} & \tfrac{1}{2}
\end{array} \right)
$$

$$
\begin{array}{l}
r_1 \to r_1 - 4r_3 \\
r_2 \to r_2 + 2r_3
\end{array}
\quad
\left( \begin{array}{ccc|ccc}
1 & 0 & 0 & 4 & -3 & -2 \\
0 & 1 & 0 & -1 & 2 & 1 \\
0 & 0 & 1 & -1 & \tfrac{1}{2} & \tfrac{1}{2}
\end{array} \right)
$$

So the inverted matrix is:

$$
\begin{bmatrix}
4 & -3 & -2 \\
-1 & 2 & 1 \\
-1 & 1/2 & 1/2
\end{bmatrix}
$$

The method used here is preferred for its overt use of matrix algebra to invert the matrix itself, and because it is straightforwardly extended to matrices with complex elements. The method will now be developed from first principles of matrix algebra. Since the method involves determinants and their evaluation, determinants will be considered first.

## DETERMINANTS: DEFINITION

A determinant is a square array formed from the elements of a square matrix.

For example, if

$$\mathbf{A} = \begin{bmatrix} a & b \\ c & d \end{bmatrix}$$

then

$$\det \mathbf{A} = \begin{vmatrix} a & b \\ c & d \end{vmatrix} = ad - bc$$

Another example, a 3rd order determinant whose matrix is

$$A = \begin{bmatrix} a_1 & b_1 & c_1 \\ a_2 & b_2 & c_2 \\ a_3 & b_3 & c_3 \end{bmatrix}$$

(1)

is

$$\begin{vmatrix} a_1 & b_1 & c_1 \\ a_2 & b_2 & c_2 \\ a_3 & b_3 & c_3 \end{vmatrix} = a_1 \begin{vmatrix} b_2 & c_2 \\ b_3 & c_3 \end{vmatrix} - b_1 \begin{vmatrix} a_2 & c_2 \\ a_3 & c_3 \end{vmatrix} + c_1 \begin{vmatrix} a_2 & b_2 \\ a_3 & b_3 \end{vmatrix}$$

$$= a_1 b_2 c_3 - a_1 c_2 b_3 - b_1 a_2 c_3 + b_1 a_3 c_2 + c_1 a_2 b_3 - c_1 a_3 b_2$$

(2).

The 2nd order determinants are called cofactors, so that

$$\begin{vmatrix} b_2 & c_2 \\ b_3 & c_3 \end{vmatrix}$$

above is the cofactor of $a_1$.

The signs of the cofactors are given by $(-1)^{i+j}$, where i and j are the ith row and jth column of its associated coefficient which is $b_1$ in the case of

$$\begin{vmatrix} a_2 & c_2 \\ a_3 & c_3 \end{vmatrix}$$

For example, the sign of

$$\begin{vmatrix} a_2 & b_2 \\ a_3 & b_3 \end{vmatrix}$$

is given by $(-1)^{1+3} = 1$ which is positive; and of

$$\begin{vmatrix} a_2 & c_2 \\ a_3 & c_3 \end{vmatrix}$$

by $(-1)^{1+2} = -1$, which is negative.

The determinant can be evaluated by 'crossing out' the row and column of the coefficient of each cofactor, which in the simple case in (1) above gives the expression in (2) which can be evaluated if the values of the elements of the determinant are known.

The properties of determinants used in this book are fully explained in Appendix 3.

## HIGHER ORDER DETERMINANTS

Expansion of 3rd order determinants gives 6 terms (triples); 4th order 24 terms, each a product of four elements, and the number of terms, each of n elements, rapidly increases with increasing n.

## CONDENSATION METHODS OF EVALUATING DETERMINANTS

One approach consists of successively reducing determinants to determinants of lower order, ending up with a 2nd order determinant which can easily be expanded. Such an approach is described in Aitken's (1967) book 'Determinants and Matrices'[1].

Following Aitken's reasoning, consider the 4th order determinant:

---

[1]Aitken, A. (1967) 'Determinants and Matrices' Oliver and Boyd Ltd.

$$|A| = \begin{vmatrix} a_1 & b_1 & c_1 & d_1 \\ a_2 & b_2 & c_2 & d_2 \\ a_3 & b_3 & c_3 & d_3 \\ a_4 & b_4 & c_4 & d_4 \end{vmatrix}.$$

assuming $a_1 \neq 0$,

first multiply rows 2 - 4 by $a_1$. This gives

$$a_1^3 |A| = \begin{vmatrix} a_1 & b_1 & c_1 & d_1 \\ a_1 a_2 & a_1 b_2 & a_1 c_2 & a_1 d_2 \\ a_1 a_3 & a_1 b_3 & a_1 c_3 & a_1 d_3 \\ a_1 a_4 & a_1 b_4 & a_1 c_4 & a_1 d_4 \end{vmatrix}$$

since multiplying a row by $\lambda$ scales $|A|$ by $\lambda$ a property of determinants.

Now another property of determinants is that subtracting a multiple of any row from any other row leaves the value of the determinant unchanged, so performing $row_2 - a_2 row_1$, $row_3 - a_3 row_1$, $row_2 - a_2 row_1$, and $row_4 - a_4 row_1$, gives

$$\begin{vmatrix} 0 & a_1 b_2 - a_2 b_1 & a_1 c_2 - a_2 c_1 & a_1 d_2 - a_2 d_1 \\ 0 & a_1 b_3 - a_3 b_1 & a_1 c_3 - a_3 c_1 & a_1 d_3 - a_3 d_1 \\ 0 & a_1 b_4 - a_4 b_1 & a_1 c_4 - a_4 c_1 & a_1 d_4 - a_4 d_1 \end{vmatrix}$$

$$= \begin{vmatrix} a_1 & b_1 & c_1 & d_1 \\ 0 & |a_1 b_2| & |a_1 c_2| & |a_1 d_2| \\ 0 & |a_1 b_3| & |a_1 c_3| & |a_1 d_3| \\ 0 & |a_1 b_4| & |a_1 c_4| & |a_1 d_4| \end{vmatrix}$$

$$= a_1 \begin{vmatrix} |a_1 b_2| & |a_1 c_2| & |a_1 d_2| \\ |a_1 b_3| & |a_1 c_3| & |a_1 d_3| \\ |a_1 b_4| & |a_1 c_4| & |a_1 d_4| \end{vmatrix}$$

$$-b_1 \begin{vmatrix} 0 & |a_1c_2| & |a_1d_2| \\ 0 & |a_1c_3| & |a_1d_3| \\ 0 & |a_1c_4| & |a_1d_4| \end{vmatrix}$$

$$+c_1 \begin{vmatrix} 0 & |a_1b_2| & |a_1d_2| \\ 0 & |a_1b_3| & |a_1d_3| \\ 0 & |a_1b_4| & |a_1d_4| \end{vmatrix}$$

$$-d_1 \begin{vmatrix} 0 & |a_1b_2| & |a_1c_2| \\ 0 & |a_1b_3| & |a_1c_3| \\ 0 & |a_1b_4| & |a_1c_4| \end{vmatrix}$$

Now the 2nd term above:

$$-b_1 \begin{vmatrix} 0 & |a_1c_2| & |a_1d_2| \\ 0 & |a_1c_3| & |a_1d_3| \\ 0 & |a_1c_4| & |a_1d_4| \end{vmatrix}$$

$$= b_1 \cdot 0 \cdot [\,|a_1c_3||a_1d_2| - |a_1c_2||a_1d_3|\,] + b_1 \cdot 0 \cdot [\cdots - \cdots] - b_1 \cdot 0 \cdot [\cdots - \cdots] = 0$$

and similarly the 3rd and 4th terms are zero. Hence we are left with the first term only.

Thus,

$$a_1^3 |A| = a_1 \begin{vmatrix} |a_1b_2| & |a_1c_2| & |a_1d_2| \\ |a_1b_3| & |a_1c_3| & |a_1d_3| \\ |a_1b_4| & |a_1c_4| & |a_1d_4| \end{vmatrix}$$

and so

$$|A| = \frac{1}{a_1^{3-1}} \begin{vmatrix} |a_1b_2| & |a_1c_2| & |a_1d_2| \\ |a_1b_3| & |a_1c_3| & |a_1d_3| \\ |a_1b_4| & |a_1c_4| & |a_1d_4| \end{vmatrix}$$

A determinant of order 4 has been condensed to one of order 3. Generalizing the method, each of the elements in the condensed determinant is of the order 2 with $a_1$ as the leading element or 'pivot', and the second element in each is in the same order as that of the original determinant, with first row and column suppressed. This is then multiplied by $1/a_1^{n-2}$, for a nth order determinant.

## THE CASE WHERE $a_1 = 0$

If $a_1 = 0$, rows or columns can be interchanged as necessary so as to bring a new row or column into position such that the leading element is non-zero (if this cannot be done, the determinant value is zero). A property of determinants is that interchanging adjacent rows or columns reverses the sign of the determinant, but its absolute value is unchanged.

## EXAMPLE

Consider the 4th order determinant:

$$|A| = \begin{vmatrix} 2 & 0 & 3 & 5 \\ 0 & 4 & -1 & 0 \\ 1 & 0 & 0 & 1 \\ 0 & 2 & 1 & 1 \end{vmatrix}.$$

Applying the condensation method,

$$|A| = \frac{1}{2^{4-2}} \begin{vmatrix} 2.4-0 & 2.-1-0 & 2.0-5.0 \\ 2.0-0 & 2.0-3 & 2.1-5.1 \\ 2.2-0 & 2.1-0 & 2.1-0 \end{vmatrix}$$

$$= \frac{1}{4} \begin{vmatrix} 8 & -2 & 0 \\ 0 & -3 & -3 \\ 4 & 2 & 2 \end{vmatrix}$$

$$= \frac{1}{4} \cdot \frac{1}{8} \begin{vmatrix} -24-0 & -24-0 \\ 16-(-8) & 16 \end{vmatrix}$$

$$= \frac{1}{32} \begin{vmatrix} -24 & -24 \\ 24 & 16 \end{vmatrix}$$

$= 1/32.[-24.16 + 24^2]$

$= 24/32.[-16 + 24]$

$= 6.$

Checking this result by the method of expansion by cofactors:

$$\begin{vmatrix} 2 & 0 & 3 & 5 \\ 0 & 4 & -1 & 0 \\ 1 & 0 & 0 & 1 \\ 0 & 2 & 1 & 1 \end{vmatrix} = 2\begin{vmatrix} 4 & -1 & 0 \\ 0 & 0 & 1 \\ 2 & 1 & 1 \end{vmatrix} - 0\begin{vmatrix} 0 & 4 & 0 \\ 1 & 0 & 1 \\ 0 & 2 & 1 \end{vmatrix} + 3\begin{vmatrix} 0 & 4 & 0 \\ 1 & 0 & 1 \\ 0 & 2 & 1 \end{vmatrix} - 5\begin{vmatrix} 0 & 4 & 0 \\ 1 & 0 & 1 \\ 0 & 2 & 1 \end{vmatrix}$$

$$= 2(4\begin{vmatrix} 0 & 1 \\ 1 & 1 \end{vmatrix} - (-1)\begin{vmatrix} 0 & 1 \\ 2 & 1 \end{vmatrix} + 0) + (3(0\begin{vmatrix} 1 & 1 \\ 0 & 1 \end{vmatrix} + 0) - 5(0\begin{vmatrix} 1 & 0 \\ 0 & 1 \end{vmatrix} - 1\begin{vmatrix} 1 & 0 \\ 0 & 2 \end{vmatrix}))$$

$= 2(-4 -2) + 3(-4) -5(-4 -2)$

$= -12 -12 + 30 = 6.$

It is apparent from this working that the

condensation method will be advantageous with determinants containing only one or two zeros.

The method of matrix inversion to be used here involves finding adjoint matrices of det **A**. Adjoint matrices will be considered next.

ADJOINT MATRICES

DEFINITION

The adjoint matrix of **A**, adj **A**, consists of the cofactors of the elements of **A**, but transposed. An element $a_{ij}$ is transposed when it is placed in the position ji.

For a square matrix of order n, there are $n^2$ elements and so there are $n^2$ cofactors. For example if

$$\mathbf{A} = \begin{vmatrix} a_1 & b_1 & c_1 \\ a_2 & b_2 & c_2 \\ a_2 & b_2 & c_2 \end{vmatrix},$$

the matrix with the cofactors of **A** as elements is

$$\begin{bmatrix} |b_2c_3| & -|a_2c_3| & |a_2b_3| \\ -|b_1c_3| & |a_1c_3| & -|a_1b_3| \\ |b_1c_2| & -|a_1c_2| & |a_1b_2| \end{bmatrix}$$

with due attention to signs.

Making element $a_{ij}$ into $a_{ji}$ gives

$$\begin{bmatrix} |b_2c_3| & -|b_1c_3| & |b_1c_2| \\ -|a_2c_3| & |a_1c_3| & -|a_1c_2| \\ |a_2b_3| & -|a_1b_3| & |a_1b_2| \end{bmatrix} = \text{adj } A$$

## THE PRODUCT A.adj A

In the above example for a 3x3 matrix, and denoting the cofactors of $a_{ij}$ by $|A_{ij}|$,

$$A \text{ adj } A = \begin{bmatrix} a_{11} & a_{12} & a_{13} \\ a_{21} & a_{22} & a_{23} \\ a_{31} & a_{32} & a_{33} \end{bmatrix} \begin{bmatrix} |A_{11}| & |A_{21}| & |A_{31}| \\ |A_{12}| & |A_{22}| & |A_{32}| \\ |A_{13}| & |A_{23}| & |A_{33}| \end{bmatrix}$$

$$= \begin{bmatrix} |A| & 0 & 0 \\ 0 & |A| & 0 \\ 0 & 0 & |A| \end{bmatrix}.$$

Since $a_{11}|A_{11}| + a_{12}|A_{12}| + a_{13}|A_{13}| = |A|$,

$a_{21}|A_{21}| + a_{22}|A_{22}| + a_{23}|A_{23}| = |A|$,

$a_{31}|A_{31}| + a_{32}|A_{32}| + a_{33}|A_{33}| = |A|$,

and by a rule of determinants expansion by alien cofactors is zero.

## AN EXPRESSION FOR THE INVERTED MATRIX IN TERMS OF |A| AND ADJ **A**

Since $\begin{bmatrix} |A| & 0 & 0 \\ 0 & |A| & 0 \\ 0 & 0 & |A| \end{bmatrix} = |A| \begin{bmatrix} 1 & 0 & 0 \\ 0 & 1 & 0 \\ 0 & 0 & 1 \end{bmatrix}$

where $\begin{bmatrix} 1 & 0 & 0 \\ 0 & 1 & 0 \\ 0 & 0 & 1 \end{bmatrix}$

is the identity matrix, **I**, and generalising, we have:

**A**.adj**A** = |A|**I**.

Or

$$\text{adj } \mathbf{A} \, \mathbf{A} = \begin{bmatrix} |A_{11}| & |A_{21}| & |A_{31}| \\ |A_{12}| & |A_{22}| & |A_{32}| \\ |A_{13}| & |A_{23}| & |A_{33}| \end{bmatrix} \begin{bmatrix} a_{11} & a_{12} & a_{13} \\ a_{21} & a_{22} & a_{23} \\ a_{31} & a_{32} & a_{33} \end{bmatrix} =$$

$$\begin{bmatrix} |A_{11}|a_{11} + |A_{21}|a_{21} + |A_{31}|a_{31} & 0 & 0 \\ 0 & |A_{12}|a_{12} + |A_{22}|a_{22} + |A_{32}|a_{32} & 0 \\ 0 & 0 & |A_{13}|a_{13} + |A_{23}|a_{23} + |A_{33}|a_{33} \end{bmatrix}$$

$$= \begin{bmatrix} |A| & 0 & 0 \\ 0 & |A| & 0 \\ 0 & 0 & |A| \end{bmatrix}$$

since only elements in the same row of adj**A** are cofactors of the elements in a column in **A**.

Hence, adj**A**.**A** produces the same result, so $|A|^{-1}$adj**A**.**A** = **I**.

But $\mathbf{A}^{-1}\mathbf{A} = \mathbf{I}$, where $\mathbf{A}^{-1}$ is the inverse of **A**.

Hence $\mathbf{A}^{-1} = |A|^{-1}.$adj**A**.

The approach described above for condensing determinants and generating adjoint matrices, which are the crux of the method used here for inverting matrices, involves taking the products of elements. If these elements are complex numbers, then in general their products will be complex numbers. We go on to consider how the method may be used in problems involving matrices with complex elements.

# METHODS OF INVERTING MATRICES WITH COMPLEX ELEMENTS

Suppose the element a + ib is multiplied by the element c + id, where a, b, c and d are real numbers. Then

(a + ib)(c + id) = ac + ibc + iad + i$^2$(bc + ad)

= ac - bd + i(bc + ad).

## MODIFYING THE EXPRESSION FOR CROSS PRODUCTS IN THE METHOD FOR REAL MATRICES

We have, for a 2 x 2 determinant:

$$\begin{vmatrix} a_{11} + ib_{11} & a_{1,j+1} + ib_{1,j+1} \\ a_{i+1,1} + ib_{i+1,1} & a_{i+1,j+1} + ib_{i+1,j+1} \end{vmatrix} =$$

$$(a_{11} + ib_{11})(a_{i+1,j+1} + ib_{i+1,j+1}) - (a_{1,j+1} + ib_{1,j+1})(a_{i+1,1} + ib_{i+1,1})$$

$$= a_{1,1}a_{i+1,j+1} + i(b_{1,1}a_{i+1,j+1} + a_{1,1}b_{i+1,j+1}) - b_{1,1}b_{i+1,j+1} -$$
$$[a_{1,j+1}a_{i+1,1} + i(b_{i,j+1}a_{i+1,1} + a_{1,j+1}b_{i+1,1}) - b_{1,j+1}b_{i+1,1}]$$

## REAL PARTS OF THE CONDENSED DETERMINANT ELEMENTS

If the elements are real only (coefficients b = 0), then the above expression reduces to

$$a_{1,1}a_{i+1,j+1} - a_{1,j+1}a_{i+1,1}.$$

If the b coefficients are not zero, then two further real-valued terms need to be added to this expression viz.:

$$-b_{1,1}b_{i+1,j+1} \text{ and } b_{1,j+1}b_{i+1,1}.$$

## IMAGINARY PARTS OF THE CONDENSED DETERMINANT ELEMENTS

From the foregoing, the expression for the imaginary part is

$$b_{1,1}a_{i+1,j+1} + a_{1,1}b_{i+1,j+1} - b_{i,j+1}a_{i+1,1} - a_{1,j+1}b_{i+1,1}.$$

## DIVISION OF COMPLEX NUMBERS

We have

$$\frac{a + ib}{c + id} = \frac{(a + ib)\cdot(c - id)}{(c + id)\cdot(c - id)} = \frac{ac + bd + i(bc - ad)}{c^2 + d^2}$$

$$= \frac{ac + cd}{c^2 + d^2} + i\frac{(bc - ad)}{c^2 + d^2} \qquad (1).$$

The method used to invert a matrix involves multiplying a condensed determinant by the factor $1/a_{11}^{n-2}$, where $a_{11}$ is the leading element of $|A|$, or $1/a_{11}^{k-1}$, where k is the order of the condensed matrix. This in turn requires each element of the condensed determinant of order k to be multiplied by the kth root of the multiplier (see p. 45), i.e.,

$$\left[\frac{1}{a_{11}^{k-1}}\right]^{1/k}$$

However, in general, the multiplier for a determinant with complex elements will itself be complex, and so the above expression involves taking a root of a complex number. A means of doing this using De Moivre's theorem is described in the next section.

## FINDING THE ROOT OF AN IMAGINARY NUMBER

By DeMoivre's theorem,

$(\cos\theta + i\sin\theta)^n = \cos n\theta + i\sin n\theta$.

Hence,

$(\cos\theta + i\sin\theta)^{1/k} =$

$\cos 1/k.\theta + i\sin 1/k.\theta$

The polar form of a complex number a + ib is

$r.(\cos\theta + i\sin\theta)$. This is illustrated graphically in Fig. 2 on the next page:

FIGURE 2: SHOWING THE POLAR REPRESENTATION OF THE COMPLEX NUMBER a + ib as r.(cosθ + isinθ)

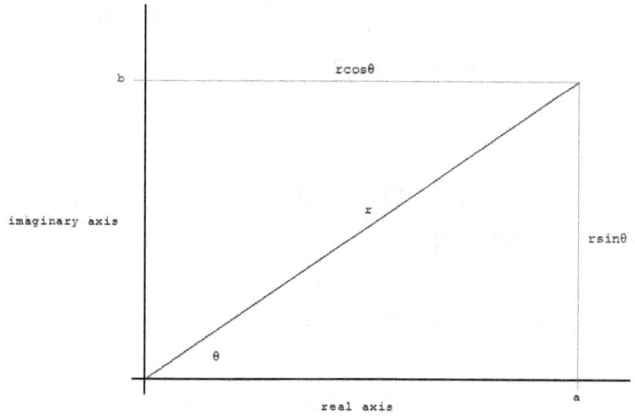

Now $\cos\theta = a/r$ and $1 - \cos^2\theta = \sin^2\theta = 1 - (a^2/r^2)$

So

$$\sin\theta = \sqrt{1 - (a^2/r^2)}$$

and $\tan\theta = \sin\theta/\cos\theta = \dfrac{\sqrt{1 - (a^2/r^2)}}{a/r}$

So

$$\theta = \tan^{-1}\left[\frac{\sqrt{1-(a^2/r^2)}}{a/r}\right]$$

Therefore, $1/k \cdot \theta = 1/k \cdot \tan^{-1}\left[\dfrac{\sqrt{1-(a^2/r^2)}}{a/r}\right] = $ Th, say

Hence, the real part of the kth root of a + ib is:

$r^{1/k} \cdot \cos$Th,

and the imaginary part is:

$r^{1/k} \cdot \sin$Th .

## APPLICATION OF THE METHOD TO A PRACTICAL PROBLEM IN ENGINEERING SCIENCE

Consider the '$\pi$' network below.

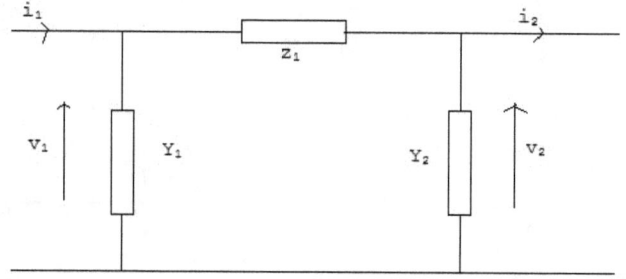

Fig 3: A '$\pi$' NETWORK CONSISTING ON AN IMPEDANCE AND TWO ADMITTANCES

Suppose $Y_1$ is $3 + 1j$, (engineers use 'j' in preference to 'i' for

$\sqrt{-1}$ since i is used for current) $z_1$ is $2 - 4j$,

that $v_1$ is $-6 - 3.8j$ and $i_1$ is $-14.4 - 19.1j$, with appropriate units.

Inserting these values into the appropriate matrices:

$$\begin{bmatrix} -6 & -3.8j \\ -14.4 & -19.1j \end{bmatrix} = \begin{bmatrix} 1+0j & 0+0j \\ 3+1j & 1+0j \end{bmatrix} \begin{bmatrix} 1+0j & 2-4j \\ 0+0j & 1+0j \end{bmatrix} \begin{bmatrix} 1+0j & 0+0j \\ 5-3j & 1+0j \end{bmatrix} \begin{bmatrix} v_2 \\ i_2 \end{bmatrix}$$

## Finding the product of the first two matrices

Let the first matrix be $A_1+jB_1$ and the second $A_2+jB_2$. Then their product $(A_1+jB_1)(A_2+jB_2)$ is:

$A_1A_2-B_1B_2+j[B_1A_2 + A_1B_2]$, where

$$A_1 = \begin{bmatrix} 1 & 0 \\ 3 & 1 \end{bmatrix} \quad A_2 = \begin{bmatrix} 1 & 2 \\ 0 & 1 \end{bmatrix}$$

$$B_1 = \begin{bmatrix} 0 & 0 \\ 1 & 0 \end{bmatrix} \quad B_2 = \begin{bmatrix} 0 & -4 \\ 0 & 0 \end{bmatrix}$$

Substituting these values into the above expression and forming the initial pairs of products gives:

$A_1A_2 - B_1B_2 + j[B_1A_2 + A_1B_2] =$

$$\begin{bmatrix} 1 & 2 \\ 3 & 7 \end{bmatrix} - \begin{bmatrix} 0 & 0 \\ 0 & -4 \end{bmatrix} + j\left( \begin{bmatrix} 0 & 0 \\ 1 & 2 \end{bmatrix} + \begin{bmatrix} 0 & -4 \\ 0 & -12 \end{bmatrix} \right)$$

$$= \begin{bmatrix} 1 & 2 \\ 3 & 11 \end{bmatrix} + j\begin{bmatrix} 0 & -4 \\ 1 & -10 \end{bmatrix}$$

(Matrix multiplication is fully explained in Appendix 2.)

## Finding the product of all three matrices

Now let

$$A_1 = \begin{bmatrix} 1 & 2 \\ 3 & 11 \end{bmatrix} \quad A_2 = \begin{bmatrix} 1 & 0 \\ 5 & 1 \end{bmatrix}$$

$$B_1 = \begin{bmatrix} 0 & -4 \\ 1 & -10 \end{bmatrix} \quad B_2 = \begin{bmatrix} 0 & 0 \\ -3 & 0 \end{bmatrix}$$

Forming the initial pairs of products gives:

$$\begin{bmatrix} 11 & 2 \\ 58 & 11 \end{bmatrix} - \begin{bmatrix} 12 & 0 \\ 30 & 0 \end{bmatrix} + j \left( \begin{bmatrix} -20 & -4 \\ -49 & 70 \end{bmatrix} + \begin{bmatrix} -6 & 0 \\ -33 & 0 \end{bmatrix} \right)$$

$$= \begin{bmatrix} -1 & 2 \\ 28 & 11 \end{bmatrix} + j \begin{bmatrix} -26 & -4 \\ -82 & -10 \end{bmatrix}$$

$$= \begin{bmatrix} -1 - 26j & 2 - 4j \\ 28 - 82j & 11 - 10j \end{bmatrix}$$

Now

$$\begin{bmatrix} v_1 \\ i_1 \end{bmatrix} = \begin{bmatrix} -1 - 26j & 2 - 4j \\ 28 - 82j & 11 - 10j \end{bmatrix} \begin{bmatrix} v_2 \\ i_2 \end{bmatrix}$$

and so,

$$\begin{bmatrix} v_2 \\ i_2 \end{bmatrix} = \begin{bmatrix} -1 - 26j & 2 - 4j \\ 28 - 82j & 11 - 10j \end{bmatrix}^{-1} \begin{bmatrix} v_1 \\ i_1 \end{bmatrix}$$

Inverting the Product Matrix

Let

$$A = \begin{bmatrix} -1 - 26j & 2 - 4j \\ 28 - 82j & 11 - 10j \end{bmatrix}$$

Then $A^{-1} = |A|^{-1} \cdot \text{adj} A$.

$$|A| = \begin{vmatrix} -1 - 26j & 2 - 4j \\ 28 - 82j & 11 - 10j \end{vmatrix}$$

From above,

$$|A| = \frac{1}{(-1 - 26j)^{2-2}} \begin{vmatrix} -1 - 26j & 11 - 10j \end{vmatrix}$$

The denominator in the first term is one and so the first term (the matrix multiplier)= 1. Forming the cross-product to evaluate the second term gives:

$$\text{Re } |A| = -1. -11 -2.28 - [(-26). -10] + [-4. -82] = 1$$

for the real part and

$$\text{Im } |A| = -26.11 + [(-1).10] - [-4.28] - 2[2. -82] = 0$$

using the expressions derived on p. 43.

Hence

$$|A|^{-1} = 1 + j0$$

### Forming the Adjoint Matrix of **A**

The co-factors of **A**, as **A** is a 2x2 matrix, are simply elements, and with due attention to signs (see Appendix 3), the matrix of co-factors is

$$\begin{bmatrix} 11-10_j & -(28-82_j) \\ -(2-4_j) & -1-26_j \end{bmatrix}$$

Transposing to form the adjoint matrix, and clearing brackets,

$$\text{Adj } \mathbf{A} = \begin{bmatrix} 11-10_j & -2+4_j \\ -28+82_j & -1-26_j \end{bmatrix}$$

and so

$$\mathbf{A}^{-1} = 1 \cdot \begin{bmatrix} 11-10_j & -2+4_j \\ -28+82_j & -1-26_j \end{bmatrix} = \begin{bmatrix} 11-10_j & -2+4_j \\ -28+82_j & -1-26_j \end{bmatrix}$$

In this case, the elements in the inverted matrix are transpositions of the elements of the original matrix (with changes of sign) because of the 1 + 0j's in the diagonals of the multiplicands.

## Forming the Product of $\mathbf{A}^{-1}$ and the input voltage/current matrix

This is

$$\begin{bmatrix} v_2 \\ i_2 \end{bmatrix} = \begin{bmatrix} 11-10j & -2+4j \\ -28+82j & -1-26j \end{bmatrix} \begin{bmatrix} -6 -3.8j \\ -14.4 -19.1j \end{bmatrix}$$

and using the procedure above, with

$$\mathbf{A}_1 = \begin{bmatrix} 11 & -2 \\ -28 & -1 \end{bmatrix} \quad \mathbf{A}_2 = \begin{bmatrix} -6 \\ -14 \end{bmatrix}$$

$$\mathbf{B}_1 = \begin{bmatrix} -10 & 4 \\ 82 & -26 \end{bmatrix} \quad \mathbf{B}_2 = \begin{bmatrix} -3.8 \\ -19.1 \end{bmatrix}$$

gives:

$$\begin{bmatrix} v_2 \\ i_2 \end{bmatrix} = \begin{bmatrix} -37.2 \\ 182.4 \end{bmatrix} - \begin{bmatrix} -38.4 \\ 185 \end{bmatrix} + j\left( \begin{bmatrix} 2.4 \\ -117.6 \end{bmatrix} + \begin{bmatrix} -3.6 \\ 125.5 \end{bmatrix} \right)$$

and so

$$\begin{bmatrix} v_2 \\ i_2 \end{bmatrix} = \begin{bmatrix} 1.2 \\ -2.6 \end{bmatrix} + j \begin{bmatrix} -1.2 \\ 182.4 \end{bmatrix} = \begin{bmatrix} 1.2 & -1.2j \\ -2.6 & +7.9j \end{bmatrix}$$

Although only 2x2 matrices were involved in this problem, the method can be used where 3x3, 4x4 and higher order matrices are involved.

## 2. SOLVING SIMULTANEOUS EQUATIONS

INTRODUCTION

Consider the circuit in Fig. 4 in where the problem is to find the unknown voltages $V_2$ - $V_5$.

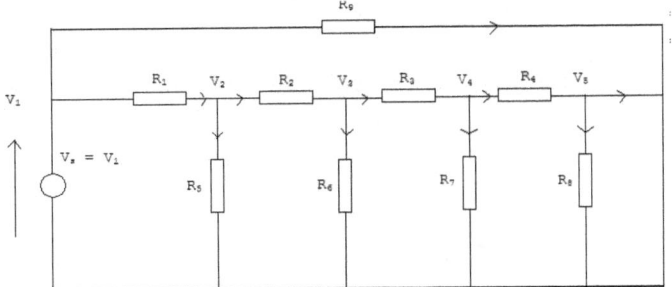

Fig. 4: A CIRCUIT CONSISTING OF RESISTANCES AND A VOLTAGE SOURCE

Using Kirchhoff's current law at junction 2:

$$\frac{V_1 - V_2}{R_1} - \frac{V_2 - V_3}{R_2} - \frac{V_2}{R_5} = 0 \qquad (1)$$

At junction 3:

$$\frac{V_2 - V_3}{R_2} - \frac{V_3 - V_4}{R_3} - \frac{V_3}{R_6} = 0 \qquad (2)$$

At junction 4:

$$\frac{V_3 - V_4}{R_3} - \frac{V_4 - V_5}{R_4} - \frac{V_4}{R_7} = 0 \qquad (3)$$

and at junction 5:

$$\frac{V_1 - V_5}{R_9} + \frac{V_4 - V_5}{R_4} - \frac{V_5}{R_8} = 0 \qquad (4)$$

From equation (1),

$V_1/R_1 - V_2/R_1 - V_2/R_2 + V_3/R_2 - V_2/R_5 = 0$

Multiplying throughout by $R_1R_2R_5$:

$R_2R_5V_1 - R_2R_5V_2 - R_1R_5V_2 + R_1R_5V_3 - R_1R_2V_2 = 0$

and so

$-(R_2R_5 + R_1R_5 + R_1R_2)V_2 + R_1R_5V_3 = -R_2R_5V_1$ (5)

From equation (2):

$V_2/R_2 - V_3/R_2 - V_3/R_3 + V_4/R_3 - V_3/R_6 = 0$,

Multiplying throughout by $R_3R_2R_6$ gives:

$$R_3R_6V_2 - R_3R_6V_3 - R_2R_6V_3 - R_2R_3V_3 + R_2R_6V_4 = 0$$

and so

$$R_3R_6V_2 - (R_3R_6 + R_2R_6 + R_2R_3)V_3 + R_2R_6V_4 = 0$$
(6).

From equation (3);

$$V_3/R_3 - V_4/R_3 - V_4/R_4 + V_5/R_4 - V_4/R_7 = 0,$$

Multiplying throughout by $R_3R_4R_7$ gives:

$$R_4R_7V_3 - R_4R_7V_4 - R_3R_7V_4 + R_7R_3V_5 - R_3R_4V_4 = 0$$

and so
$$R_4R_7V_3 - (R_4R_7 + R_3R_7 + R_4R_3)V_3 + R_3R_7V_5 = 0$$
(7).

Finally, from equation (4):
$$V_1/R_9 - V_5/R_9 + V_4/R_4 - V_5/R_4 - V_5/R_8 = 0.$$
Multiplying throughout by $R_8R_4R_9$ gives:

$R_4R_8V_1 - R_4R_8V_5 + R_8R_9V_4 - R_8R_9V_5 - R_8R_4V_5 = 0$

and so

$R_4R_8V_1 - (R_4R_8 + R_8R_9 + R_4R_8)V_5 + R_4R_9V_4 = 0$
(8).

If $V_1 = V_s$ is known, we have 4 simultaneous equations in 4 unknowns. If $R_1 - R_9$ are known, we have four equations of the form:

$au_1 + bu_2 + cu_3 + du_4 = $ constant,

where $u_1, u_2, u_3$ and $u_4$ represent the unknowns, here the voltages $V_2 - V_5$, and a, b, c, and d the coefficients of the unknowns, which in ths case are products of resistance values and some may be zero.

## Solving Simultaneous Equations by the Gauss-Jordan Method

One popular method of solution is known as Gauss-Jordan elimination, which is performed below on a 4th order set of simultaneous equations.

Suppose we have
$r_1: 2w + x + y + 2z = 15$
$r_2: w + 2x + 2y + 3z = 23$
$r_3: 3w - 2x - 3y + z = -6$
and
$r_4: -2w + 3x - y - z = -3,$
where $r_1$ - $r_4$ denote the four rows.

The aim is to use each row in turn to eliminate one variable from the remaining rows as follows:
$r_1: 2w + x + y + 2z = 15$
$r_2 \rightarrow$ **$2r_2 - r_1$**: $0w + 3x + 3y + 4z = 31$
$r_3 \rightarrow$ **$r_3 - 3/2.r_1$**: $0w - 7/2.x - 9/2.y - 2z = -57/2$
$r_4 \rightarrow$ **$r_4 + r_1$**: $0w + 4x - 0y + z = 12;$

$r_1 \rightarrow$ **$3r_1 - r_2$**: $6w + 0x + 0y + 2z = 14$
$r_2: 0w + 3x + 3y + 4z = 31$
$r_3 \rightarrow$ **$r_3 - 7/6.r_2$**: $0w + 0x - y + 8/3.z = 23/3$
$r_4 \rightarrow$ **$r_4 - 4/3.r_2$**: $0w + 0x - 4y - 13/3.z = -88/3;$
$r_4 \rightarrow$ **$r_4 - 4r_3$**: $0w + 0x + 0y - 15z = -60$
so $z = 4;$

$r_3 \rightarrow$ **$45/8.r_3 + r_4$**: $0w + 0x - 45/8.y + 0z = -35/8$
so $y = 3;$

$r_1 \rightarrow$ **$15/2.r_1 + r_4$**: $45w + 0x + 0y + 0z = -88/3$
so $w = 1;$

$r_2 \to 15/4.r_2 + r_3$: $0w + 45/4.x + 45/4y + 0z = 225/4$
$r_2 \to r_2 + 2r_3$: $0w + 45/4.x + 0y + 0z = 45/2$;
so $x = 2$.

Even though for the purposes of illustration simple values of coefficients were chosen, the author made several mistakes in the arithmetical calculations which increased the labour involved in finding a solution. Also, since the solution set was chosen before hand, errors were easier to locate. With more realistically valued coefficients (which it will be remembered are products of resistance values in the circuit example), greater likelihood of error results from the more awkward fractions involved.

Another method of solving simultaneous equations is a matrix one, which follows naturally from the work on inverting matrices described previously.

# SOLVING SIMULTANEOUS EQUATIONS USING ADJOINT MATRICES AND DETERMINANTS

From the previous work, the inverse of a matrix is given by

$A^{-1} = |A|^{-1}.\mathrm{adj}A$.

Let **A** be the matrix of coefficients of the unknowns, **x** the column matrix of unknowns and **k** the column matrix of r.h.s. constants.

Then the set of equations is $Ax = k$

and the solution is given by $x = A^{-1}k$

In the case of a set of 4 equations in four unknowns $x_1, x_2, x_3, x_4$,

$$\begin{bmatrix} x_1 \\ x_2 \\ x_3 \\ x_4 \end{bmatrix} = |A^{-1}| \begin{bmatrix} |A_{11}| & |A_{21}| & |A_{31}| & |A_{41}| \\ |A_{12}| & |A_{22}| & |A_{32}| & |A_{42}| \\ |A_{13}| & |A_{23}| & |A_{33}| & |A_{43}| \\ |A_{14}| & |A_{24}| & |A_{34}| & |A_{44}| \end{bmatrix} \begin{bmatrix} k_1 \\ k_2 \\ k_3 \\ k_4 \end{bmatrix}$$

(1)

where

$$A = \begin{bmatrix} a_{11} & a_{12} & a_{13} & a_{14} \\ a_{21} & a_{22} & a_{23} & a_{24} \\ a_{31} & a_{32} & a_{33} & a_{34} \\ a_{41} & a_{42} & a_{43} & a_{44} \end{bmatrix}$$

But the matrix product in the r.h.s. is the product of each k and the cofactor of the corresponding element in $|A|$:

$$|A_{11}|k_1 \quad |A_{21}|k_2 \quad |A_{31}|k_3 \quad |A_{41}|k_4 \; = \; \begin{vmatrix} k_1 & a_{12} & a_{13} & a_{14} \\ k_2 & a_{22} & a_{23} & a_{24} \\ k_3 & a_{32} & a_{33} & a_{34} \\ k_4 & a_{42} & a_{43} & a_{44} \end{vmatrix}$$

$$|A_{12}|k_1 \ |A_{22}|k_2 \ |A_{32}|k_3 \ |A_{42}|k_4 \ = \ \begin{vmatrix} a_{11} & k_1 & a_{13} & a_{14} \\ a_{21} & k_2 & a_{23} & a_{24} \\ a_{31} & k_3 & a_{33} & a_{34} \\ a_{41} & k_4 & a_{43} & a_{44} \end{vmatrix}$$

$$|A_{13}|k_1 \ |A_{23}|x_2 \ |A_{33}|k_3 \ |A_{43}|k_4 \ = \ \begin{vmatrix} a_{11} & a_{12} & k_1 & a_{14} \\ a_{21} & a_{22} & k_2 & a_{24} \\ a_{31} & a_{32} & k_3 & a_{34} \\ a_{41} & a_{42} & k_4 & a_{44} \end{vmatrix}$$

$$|A_{14}|k_1 \ |A_{24}|k_2 \ |A_{34}|k_3 \ |A_{44}|k_4 \ = \ \begin{vmatrix} a_{11} & a_{12} & a_{13} & k_1 \\ a_{21} & a_{22} & a_{23} & k_2 \\ a_{31} & a_{32} & a_{33} & k_3 \\ a_{41} & a_{42} & a_{43} & k_4 \end{vmatrix}$$

Substituting these equations expression on the r.h.s. of equation (1) we therefore have:

$$x_1 = \frac{\begin{vmatrix} k_1 & a_{12} & a_{13} & a_{14} \\ k_2 & a_{22} & a_{23} & a_{24} \\ k_3 & a_{32} & a_{33} & a_{34} \\ k_4 & a_{42} & a_{43} & a_{44} \end{vmatrix}}{\begin{vmatrix} a_{11} & a_{12} & a_{13} & a_{14} \\ a_{21} & a_{22} & a_{23} & a_{24} \\ a_{31} & a_{32} & a_{33} & a_{34} \\ a_{41} & a_{42} & a_{43} & a_{44} \end{vmatrix}}$$

$$x_2 = \frac{\begin{vmatrix} a_{11} & k_1 & a_{13} & a_{14} \\ a_{21} & k_2 & a_{23} & a_{24} \\ a_{31} & k_3 & a_{33} & a_{34} \\ a_{41} & k_4 & a_{43} & a_{44} \end{vmatrix}}{\begin{vmatrix} a_{11} & a_{12} & a_{13} & a_{14} \\ a_{21} & a_{22} & a_{23} & a_{24} \\ a_{31} & a_{32} & a_{33} & a_{34} \\ a_{41} & a_{42} & a_{43} & a_{44} \end{vmatrix}}$$

$$x_3 = \frac{\begin{vmatrix} a_{11} & a_{12} & k_1 & a_{14} \\ a_{21} & a_{22} & k_2 & a_{24} \\ a_{31} & a_{32} & k_3 & a_{34} \\ a_{41} & a_{42} & k_4 & a_{44} \end{vmatrix}}{\begin{vmatrix} a_{11} & a_{12} & a_{13} & a_{14} \\ a_{21} & a_{22} & a_{23} & a_{24} \\ a_{31} & a_{32} & a_{33} & a_{34} \\ a_{41} & a_{42} & a_{43} & a_{44} \end{vmatrix}}$$

and

$$x_4 = \frac{\begin{vmatrix} a_{11} & a_{12} & a_{13} & k_1 \\ a_{21} & a_{22} & a_{23} & k_2 \\ a_{31} & a_{32} & a_{33} & k_3 \\ a_{41} & a_{42} & a_{43} & k_4 \end{vmatrix}}{\begin{vmatrix} a_{11} & a_{12} & a_{13} & a_{14} \\ a_{21} & a_{22} & a_{23} & a_{24} \\ a_{31} & a_{32} & a_{33} & a_{34} \\ a_{41} & a_{42} & a_{43} & a_{44} \end{vmatrix}}$$

## APPLICATION OF THE METHOD TO A PRACTICAL PROBLEM IN ENGINEERING SCIENCE

Consider the circuit in Fig. 5:

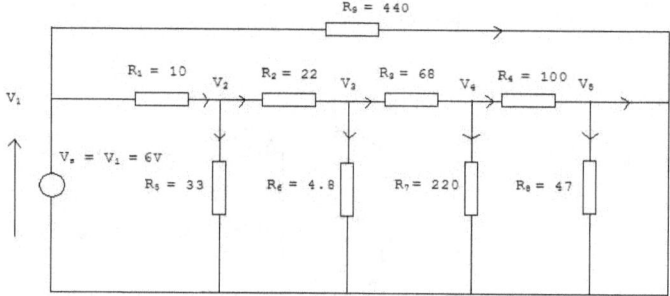

FIG. 5 : A CIRCUIT CONTAINING RESISTANCES AND A VOLTAGE SOURCE

Using equations (5) - (8) above the equations to be solved are:

$-(22 \times 33 + 10 \times 33 + 10 \times 22).v_2 + 10 \times 33.v_3 = -22 \times 33 \times 6$

$68 \times 4.8.v_2 - (68 \times 4.8 + 22 \times 4.8 + 22 \times 68).v_3 + 22 \times 4.8.v_4 = 0$

$100 \times 22.v_3 - (100 \times 220 + 68 \times 220 + 68 \times 100).v_4 + 68 \times 220.v_5 = 0$

$47 \times 440.v_4 - (100 \times 47 + 47.440 + 100.440).v_5 = -100 \times 47 \times 6$

Simplifying and rearranging gives:
$-1276v_2 + 330v_3 = -4356$
$326.4v_2 - 1928v_3 + 105.6v_4 = 0$
$22000v_3 - 43760v_4 + 14960v_5 = 0$
$20680v_4 - 69380v_5 = -28200.$

From p. 66f, the matrix solutions to these equations are:

$$v_2 = \frac{\begin{vmatrix} -4356 & 330 & 0 & 0 \\ 0 & -1928 & 105.6 & 0 \\ 0 & 22000 & -4370 & 14960 \\ -2820 & 0 & 20680 & -69380 \end{vmatrix}}{\begin{vmatrix} -1276 & 330 & 0 & 0 \\ 326.4 & -1928 & 105.6 & 0 \\ 0 & 22000 & -43760 & 14960 \\ 0 & 0 & 20680 & -69380 \end{vmatrix}}$$

$$v_3 = \cfrac{\begin{vmatrix} -1276 & -4356 & 0 & 0 \\ 326.4 & 0 & 105.6 & 0 \\ 0 & 0 & -43760 & 14960 \\ 0 & -28200 & 20680 & -69380 \end{vmatrix}}{\begin{vmatrix} -1276 & 330 & 0 & 0 \\ 326.4 & -1928 & 105.6 & 0 \\ 0 & 22000 & -43760 & 14960 \\ 0 & 0 & 20680 & -69380 \end{vmatrix}}$$

$$v_4 = \frac{\begin{vmatrix} -1276 & 330 & -4356 & 0 \\ 326.4 & -1928 & 0 & 0 \\ 0 & 22000 & 0 & 14960 \\ 0 & 0 & -28200 & -69380 \end{vmatrix}}{\begin{vmatrix} -1276 & 330 & 0 & 0 \\ 326.4 & -1928 & 105.6 & 0 \\ 0 & 22000 & -43760 & 14960 \\ 0 & 0 & 20680 & -69380 \end{vmatrix}}$$

and

$$v_5 = \frac{\begin{vmatrix} -1276 & 330 & 0 & -4356 \\ 326.4 & -1928 & 105.6 & 0 \\ 0 & 22000 & -43760 & 0 \\ 0 & 0 & 20680 & -28200 \end{vmatrix}}{\begin{vmatrix} -1276 & 330 & 0 & 0 \\ 326.4 & -1928 & 105.6 & 0 \\ 0 & 22000 & -43760 & 14960 \\ 0 & 0 & 20680 & -69380 \end{vmatrix}}$$

The steps in the calculation of $v_2$ will be performed below; the calculation of the remaining voltages following a similar procedure is left as an exercise for the reader.

Calculation of $v_2$

Using the condensation method described earlier, the numerator is equal to:

$$\frac{1}{(-4356)^{4-2}} \begin{vmatrix} -4356.-1928 & -4356.105.6 & 0 \\ -4356.22000 & -4356.-43760 & -4356.14960 \\ -(330.-2820) & -4356.20680 & -4356.-69380 \end{vmatrix}$$

$$= \frac{1}{(-4356)^{4-2}} \begin{vmatrix} 8398368 & -459993.6 & 0 \\ -95832000 & 190618560 & -65165760 \\ 930600 & -90082080 & 302219280 \end{vmatrix}$$

$$= \frac{1}{(-4356)^{4-2}} \cdot \frac{1}{8398368^{4-3}} \begin{vmatrix} 1.556802708\,E15 & -5.472860335\,E14 \\ -7.56114388\,E14 & 2.53814873\,E15 \end{vmatrix}$$

$$= \frac{5.27016555\,E-8}{8398368} \cdot 3.537585972\,E30 = 2.21991507\,E16$$

The denominator is equal to:

$$\frac{1}{(-1276)^{4-2}} \begin{vmatrix} -1276.-1928 - 330.326.4 & -1276.105.6 - 0.326.4 & -1276.0 - 0.326.4 \\ -1276.22000 - 330.0 & -1276.-43760 - 0.0 & -1276.14960 - 0.0 \\ -1276.0 - 330.0 & -1276.20680 - 0.0 & -1276.-69380 - 0.0 \end{vmatrix}$$

$$= \frac{1}{(-1276)^{4-2}} \begin{vmatrix} 2352416 & -134745.6 & 0 \\ -28072000 & 55837760 & -19088960 \\ 0 & -26387680 & 88528880 \end{vmatrix}$$

$$= \frac{1}{(-1276)^{4-2}} \cdot \frac{1}{2352416} \begin{vmatrix} 2352416 \cdot 55837760 - (-134745.6 \cdot -28072000) & 2352416 \cdot -19088960 - 0 \cdot -28072000 \\ -2352416 \cdot -26387680 \cdot (-13745.6.0) & 2352416 \cdot 88528880 - 0.0 \end{vmatrix}$$

$$= \frac{6.14184216\ E\text{-}7}{2352416} \begin{vmatrix} 1.275710615\ E14 & -4.490517493\ E13 \\ -6.207480063\ E13 & 2.082567538\ E14 \end{vmatrix}$$

$$= \frac{6.14184216\ E\text{-}7}{2352416} \cdot 2.378125857\ E28 = 6.20896673\ E15$$

Hence,

$$v_2 = \frac{2.21991507\ E16}{6.20896673\ E15} = 3.6 \text{ volt}$$

Even though full accuracy has been kept in all the steps in calculation, the result has been quoted to the 1 d.p. required in practice as in the case of the determination of the output voltage/current in the previous section looking at transmission voltages.

# 3. SOLVING POLYNOMIAL EQUATIONS

## INTRODUCTION

### THE TRANSFER CHARACTERISTIC OF A D.C. AMPLIFIER

Consider the transfer (input-output) characteristic of a D.C. amplifier shown below:

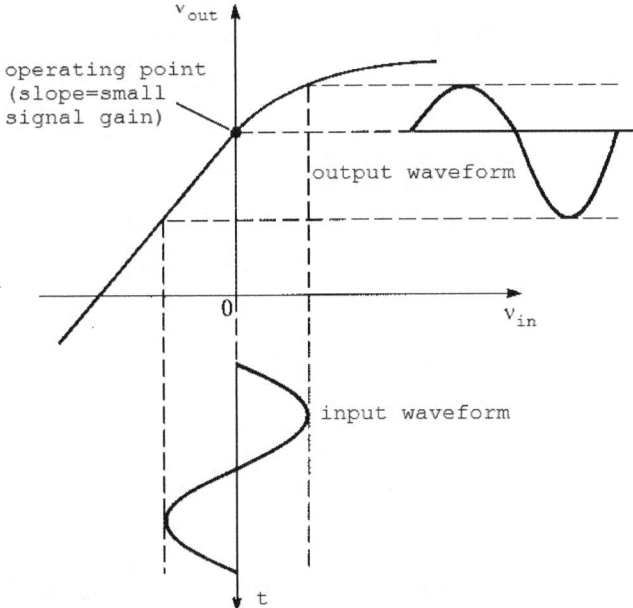

FIG 6 : THE TRANSFER CHARACTERISTIC OF A D.C. AMPLIFIER

This can be represented by a power series of the type:

$$v_0 = a + bv + cv^2 + dv^3 + ev^4 + ...,$$

where $v_0$ is the output voltage and $v$ the input voltage.

This equation can be rearranged to be in the form of the polynomial equation $f(v) = 0$ where $f(v) = \text{constant} + a_1v + a_2v^2 + a_3v^3 + a_4v^4 + ...$
with constant $= (a - v_0)$, $a_1 = b$, $a_2 = c$, $a_3 = d$ and $a_4 = e$.

The question arises, given appropriate values of the coefficients of $v_n$, what is the minimum value of input voltage $v$ which will produce a given output voltage $v_0$?

SOLVING POLYNOMIAL EQUATIONS

For polynomial equations in two unknowns (quadratics) the formula ( 'x' is used as a variable instead of v in the general argument which follows):

$$x = \frac{-b \pm \sqrt{b^2 - 4ac}}{2a}$$

can be used to solve it. A more complex formula can be used for polynomial equations of order 3 (cubics) and a still more complex formula for polynomial equations of order 4 (quartics). There are no formulae for higher-order polynomial equations and so iterative approximation methods have to be resorted to solve these equations.

## A METHOD OF SOLVING POLYNOMIAL EQUATIONS USING THE NEWTON-RAPHSON APPROACH

Suppose that the polynomial equation $y = f(x)$ has a solution (i.e. for which $f(x) = 0$) of $x = a$ and the actual solution lies between $x = a$ and $x = b$. The Taylor series about a for f is:
$f(a) + f'(a)(x - a) + f''(a)(x - a)^2/2! + ...$

If $x = a + h$,

$f(a + h) = f(a) + f'(a)(a + h - a) + f''(a))(a + h - a)^2/2! + ...$
$= f(a) + f'(a)h + f''(a)h2/2! + ...$

If h is small,
$0 = f(a + h) \approx f(a) + f'(a)h$,

and so $h \approx -f(a)/f'(a)$.
So a second approximation is equal to:

$$a - f(a)/f'(a) \qquad (1).$$

If this process is repeated, closer and closer approximations to a solution can be obtained. This is the essence of the Newton Raphson approach. In the case of polynomials, finding $f'(x)$ is straight forward because every term in a polynomial $f(x)$ is of the form $a_k x^k$, the first derivative of which is simply $a^k.k.x^{k-1}$, and differentiating a polynomial of order n therefore gives rise to another polynomial of order n - 1.

(1) finds the first root, but subsequent roots cannot be found simply by repeating the procedure, because the same root would be returned. This problem is addressed in the next section.

## FINDING SUBSEQUENT SOLUTIONS

Having found one solution, the problem arises of how to find the others. Repeating the iteration procedure with the same polynomial would just lead to the same solution being found. The answer to the problem is to 'take out' the previously found root of the inputted polynomial by dividing it through by (x - a), where a is the previously found root.

Suppose a, a', a'', a''', $a^{iv}$, ... are the roots of $f_n(x)=0$.

Then

$$f_n(x) = a_n(x - a)(x - a')(x - a'')(x - a''')(x - a^{iv})...$$

Dividing the r.h.s. through by (x - a) gives

$$f_{n-1}(x) = a_n(x - a')(x - a'')(x - a''')(x - a^{iv})...$$

and so a is no longer a root (unless it is a repeated root) and if the iterative procedure is repeated with $f_{n-1}(x)$ a new root will be found (given that equation (1) converges).

## EXAMPLE

Consider the polynomial equation

$3x^3 - 12x^2 + 3x + 18 = 0$.

Suppose further than one solution found is $x = 3$.
Dividing the l.h.s. of the above equation by $x-3$:

$$
\begin{array}{r}
3x^2 - 3x - 6 \phantom{xxxx} \\
x-3 \overline{\smash{\big)}\, 3x^3 - 12x^2 + 3x + 18} \\
\underline{3x^3 - 9x^2} \phantom{xxxxxxxx} \\
-3x^2 + 3x + 18 \\
\underline{-3x^2 + 9x} \phantom{xxxx} \\
-6x + 18 \\
\underline{-6x + 18} \\
\end{array}
$$

And repeated application of the method to $3x^2 - 3x - 6 = 0$ will yield a new solution.

## INITIAL VALUE OF THE APPROXIMATION

It is not necessary to start a value of a which is close to a solution of $f(x) = 0$. However, as will be explained when we come on to solve a real problem in engineering science involving the solution of a polynomial equation, the closer the starting value to a solution in general the fewer the iterations will be required to obtain a solution of a given accuracy.

Of course, polynomial equations may have real, imaginary or a mix of real and imaginary solutions, and a procedure will be described which finds both (with the caveat mentioned above).

## FINDING COMPLEX SOLUTIONS
## COMPLEX CONJUGATE ROOTS

Since the coefficients $a_i$ of $x_i$ are real, complex roots come in 'conjugate pairs'. Thus, if $a = b + ic$ is a root, then so is $b - ic$.

To find complex roots, the first approximation must be a complex number and in practice, where one has little idea of a solution, after some trial and error, using $1 + 0.99i$ was found to work satisfactorily, in that it enabled complex solutions to most polynomial equations to be found.

Since a is now a complex number, the expression in equation (1) above has the form:

$$a - f(a)/f'(a) = c + id - (e + if)/(g + ih),$$

where $a = c + id$, $f(a) = e + if$ and $f'(a) = g + ih$.

This is equal to:

$$c + id - (e + if)(g - ih)/(g^2 + h^2)$$
$$= c + id - [(eg + fh)/(g^2 + h^2) + (fg - eh)/(g^2 + h^2)i]$$
$$= [c - eg + fh)/(g^2 + h^2)] + i[d - fg - eh)/(g^2 + h^2)]$$

## FINDING SUCCESSIVE COMPLEX SOLUTIONS

Having found a complex root of the original polynomial and therefore another which is its complex conjugate, the polynomial has to be divided by $(x - [a + ib])(x - [a - ib])$ after the manner described above for real roots, so that, effectively, this root is taken out of the resulting quotient (i.e. the next lower order polynomial).

## APPLICATION OF THE METHOD TO A PRACTICAL PROBLEM IN ENGINEERING SCIENCE

Referring back to the situation in Fig. 6, consider the 'push-pull' output stage in Fig. 7:

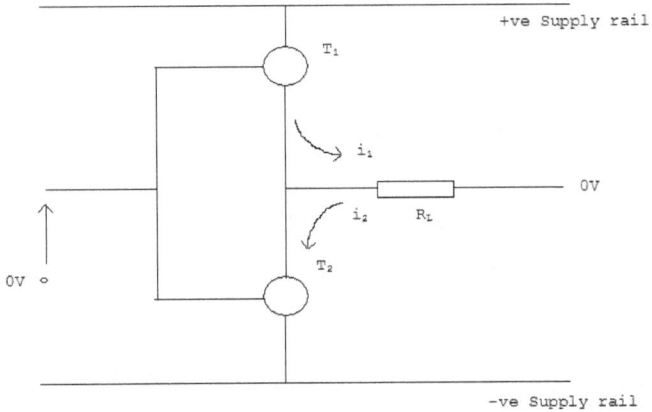

**FIG 7 CIRCUIT DIAGRAM OF A PUSH-PULL AMPLIFIER**

The two transistors shown are matched n-p-n and p-n-p types. A positive input signal switches T1 to conduct, 'pushing' current into the load. A negative input signal causes T2 to conduct, 'pulling' current through the load. The transfer characteristic for T1 is:

$$i_1 = a + bv + cv^2 + dv^3 + ev^4 + fv^5...,$$ subsequent terms being negligible

For T2 it is:
$$i_2 = a + b(-v) + c(-v)^2 + d(-v)^3 + e(-v)^4 + f(-v)^5 + ...$$

The resulting load current $i_L$ is therefore:

$i_L = i_1 - i_2 = 2bv + 2dv^3 + 2fv^5$, neglecting higher order terms, and $v_0 = V_L = i_L R_L$.

Suppose that b = 10, d = 1 and f = 0.05 and that the maximum value of $i_L$ required is 3A. If $R_L$ is 15Ω, what is the minimum input voltage amplitude v given the maximum load current?

We have

$V_L = v_0 = 3 \times 15 = 20v + 2v^3 + 0.1v^5$.

Setting up the polynomial equation f(v) = 0 gives:
$0 = -45 + 20v + 2v^3 + 0.1v^5$.

In general, a polynomial equation of order n has n solutions (or 'roots'). Here, we are interested only in real solutions, and so the initial value of v can be set at some real value- and the iteration procedure will then return only real solutions. Furthermore, we can reduce the labour involved in the calculation by reducing the number of iterations required if an approximate value for a solution can be determined at the outset. So, in this case, it is worth drawing a rough graph of f(v) to see how it behaves close to the v axis. Note also that all the terms in v and higher powers of v are positive, and the powers themselves are odd numbers. This means that the function will be symmetrical about f(v)=-45, where v=0, and accelerating in slope. All negative values of v produce negative values of f(v), but as v becomes increasingly positive, the sum of terms in v and higher powers of v overhaul the negative value of the constant -45 at some value of v, and the curve of f(v) crosses the v axis once and remains positive for all larger values of v. A sketch based on these characteristics is shown below:

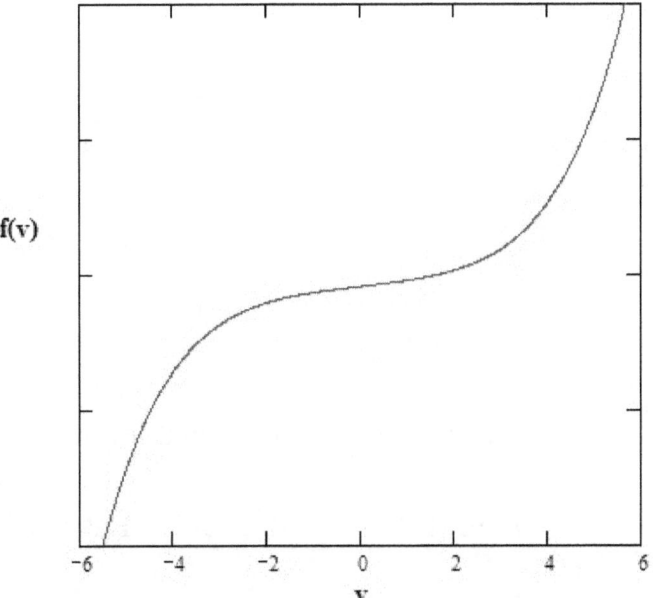

Fig. 8: A rough sketch of f(v) vs v

It can be seen that, although a polynomial of order 5 will have 5 roots, this one has only one real root and so the other four, 2 pairs of conjugates (which are of no interest in this particular example), must be complex.

The curve appears to cross the v axis somewhere between v=1.5 and v=2, so a good 1st approximation and initial value with which to start the iterative process is with v=1.75.

**Performing Successive iterations on f(v)**

The second approximation is given by:

$$v - \frac{f(v)}{f'(v)} = 1.75 - \frac{-45 + 20v + 2v^3 + 0.1v^5}{20 + 6v^2 + 0.5v^4}\bigg|_{v=1.75}$$

= 1.75 - (-0.054802938) = 1.695197061

The third approximation is obtained by substituting this new value of v into the above expression, which gives the third approximation:

1.695197061 -0.001131367 = 1.694065694

Similarly, the 4th approximation is

1.694065694 - (0.000000465) = 1.694065228

Since the 4th and third approximations agree to one decimal place accuracy when rounded up we need proceed no further, and the required solution is v=1.7 volt, and so in answer to the question posed v = 1.7 volt is the minimum input voltage which produces a load current of 3A in a 15Ω load.

## CONCLUSION

Three mathematical methods in solving practical problems in Engineering Science have been described.

The first is a method to invert matrices with complex elements which finds the inverted matrix by dividing the adjoint matrix of the inputted matrix by the value of its determinant.

When used with a transmission matrix approach to solving A.C. circuits for currents and voltages, it was shown to provide a less laborious method than using current and voltage laws. Although the method was illustrated with 2x2 matrices, it can be used to invert higher-order matrices. In fact, the author has used it successfully to invert 10x10 matrices with complex elements using a computer program based on the method.

The second method solves higher-order simultaneous equations. The mathematical approach on which the first (matrix) method is based was shown to produce the solution set in the form of quotients, with the value of the determinant containing the coefficients of the unknowns in each equation as rows in the divisor. The dividend is the same determinant only with the column of r.h.s. constants in the equations substituted for the corresponding column. In solving D.C. circuits using Kirchhoff's laws, a fairly complex circuit of resistors and voltage source can give rise to 4 or more unknown voltages giving four or more equations in four or more unknowns to solve. Such equations can be solved by 'row reduction' methods, though tedious and error-prone. The alternative matrix method described comes up with the required solution voltages once the required coefficient of each unknown voltage is calculated from the circuit resistance values as shown earlier.

The third method solves polynomial equations of the form

$$0 = \text{constant} + a_1 x + a_2 x^2 + a_3 x^3 + a_4 x^4 + \ldots$$

using an iterative, successive approximation approach based on Newton-Raphson's approach and was shown to be especially useful with polynomial equations of order greater than 2, for which formulae are either complex or non-existent. The method was applied to the transfer characteristic of a push-pull amplifier output stage to find a minimum voltage which would produce the required load current. When the polynomial equation has complex solutions, the method finds these also (in most cases).

## APPENDIX 1: FINDING THE FORM OF TRANSMISSION MATRIX FOR CIRCUIT STAGES CONTAINING AN IMPEDANCE, A RESISTANCE, AN ADMITTANCE AND A CONDUCTANCE

<u>1. Impedance</u>

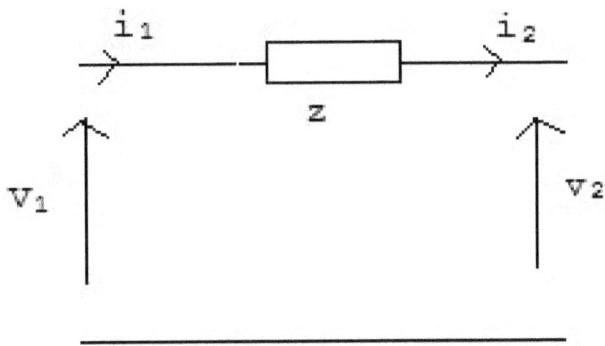

The from of the matrix equation for the above circuit stage is:

$$\begin{bmatrix} a & b \\ c & d \end{bmatrix} \begin{bmatrix} v_1 \\ i_1 \end{bmatrix} = \begin{bmatrix} v_2 \\ i_2 \end{bmatrix}$$

Now $i_1 = i_2$

so

$c.v_1 + d i_1 = i_2 = i_1$

Equating coefficients of $v_1$ and $i_1$ on each side of this equation gives

$c = 0, d = 1$.

Also, $a v_1 + b i_1 = v_2$           (1)

But $i_1 = (v_2 - v_1)/z$ and so

$z.i_1 + v_1 = v_2$. Similarly, (1) gives:
$a v_1 + b i_1 = v_1 + z.i_1$.

Equating coefficients again, $a = 1, b = z$.

Hence the transmission matrix for an impedance is

$$\begin{bmatrix} 1 & z \\ 0 & 1 \end{bmatrix}$$

## 2. Resistance

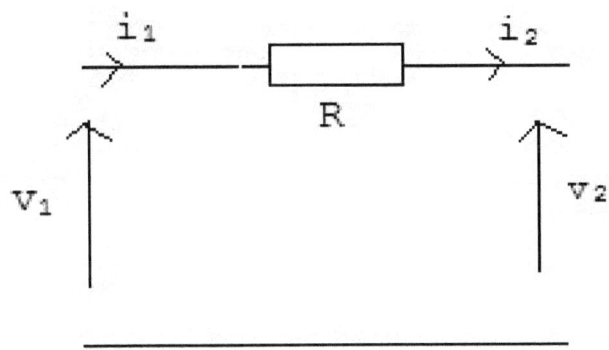

We have

$$\begin{bmatrix} e & f \\ g & h \end{bmatrix} \begin{bmatrix} v_1 \\ i_1 \end{bmatrix} = \begin{bmatrix} v_2 \\ i_2 \end{bmatrix}$$

and so

$e.v_1 + f i_1 = i_2 = i_1 g.v_1 + h i_1 = v_2$
$i_2 = (v_2 - v_1)/R.$

These equations are the same as those for the impedance with R replacing z and so the required transmission matrix is:

$$\begin{bmatrix} 1 & R \\ 0 & 1 \end{bmatrix}$$

## 3. Admittance

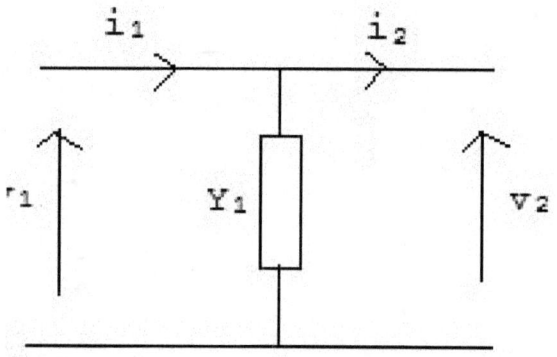

$$\begin{bmatrix} k & l \\ m & n \end{bmatrix} \begin{bmatrix} v_1 \\ i_1 \end{bmatrix} = \begin{bmatrix} v_2 \\ i_2 \end{bmatrix}$$

so
$k.v_1 + li_1 = v_2 = v_1$ (2)

$$m.v_1 + ni_1 = i_2 \qquad (3).$$

Equating coefficients in (2) gives $k = 1, l = 0$.

Now $i_2 = i_1 + v_2 Y$. Substituting this in (3) gives:

$$m.v_1 + ni_1 = v_2 Y + i_1 = v_1 Y + i_1 .$$

Equating coefficients again. $m = Y, n = 1$.
So the transmission matrix for an admittance is:

$$\begin{bmatrix} 1 & 0 \\ Y & 1 \end{bmatrix}$$

### 4. conductance
By similar reasoning to that in 2. above, the transmission matrix for a conductance is:

$$\begin{bmatrix} 1 & 0 \\ G & 1 \end{bmatrix}$$

# APPENDIX 2
## Matrix Multiplication

Consider the matrices:

$$A = \begin{bmatrix} a_{11} & a_{12} & a_{13} & \cdots & a_{1n} \\ a_{21} & a_{22} & a_{23} & \cdots & a_{2n} \\ a_{31} & a_{32} & a_{33} & \cdots & a_{3n} \\ \cdot & \cdot & \cdot & \cdots & \cdot \\ \cdot & \cdot & \cdot & \cdots & \cdot \\ \cdot & \cdot & \cdot & \cdots & \cdot \\ a_{n1} & a_{n2} & a_{n3} & \cdots & a_{nn} \end{bmatrix}$$

and

$$B = \begin{bmatrix} b_{11} & b_{12} & b_{13} & \cdots & b_{1n} \\ b_{21} & b_{22} & b_{23} & \cdots & b_{2n} \\ b_{31} & b_{32} & b_{33} & \cdots & b_{3n} \\ \cdot & \cdot & \cdot & \cdots & \cdot \\ \cdot & \cdot & \cdot & \cdots & \cdot \\ \cdot & \cdot & \cdot & \cdots & \cdot \\ b_{n1} & b_{n2} & b_{n3} & \cdots & b_{nn} \end{bmatrix}$$

Then denoting the element in the ith row and jth column of **AB** by $AB_{ij}$,

$$AB_{ij} = \sum_{k=1}^{n} a_{ik} b_{kj}$$

if **B** is a column matrix with elements $b_{i1}$, then

$$AB_{ij} = \sum_{k=1}^{n} a_{ik} b_{k1}$$

EXAMPLE

$$\mathbf{A} = \begin{bmatrix} 2 & 1 \\ 3 & 4 \end{bmatrix} \qquad \mathbf{B} = \begin{bmatrix} 5 & 7 \\ 8 & 9 \end{bmatrix}$$

$$AB_{11} = \sum_{k=1}^{2} a_{1k} b_{k1} = a_{11} b_{11} + a_{12} b_{21} = 2 \times 5 + 1 \times 8 = 18$$

$$AB_{12} = \sum_{k=1}^{2} a_{1k} b_{k2} = a_{11} b_{12} + a_{12} b_{22} = 2 \times 7 + 1 \times 9 = 23$$

$$AB_{21} = \sum_{k=1}^{2} a_{2k} b_{k1} = a_{21} b_{11} + a_{22} b_{21} = 3 \times 5 + 4 \times 8 = 47$$

and

$$AB_{22} = \sum_{k=1}^{2} a_{2k} b_{k2} = a_{21} b_{12} + a_{22} b_{22} = 3 \times 7 + 4 \times 9 = 57$$

So

$$\mathbf{A}\,\mathbf{B} = \begin{bmatrix} 18 & 23 \\ 47 & 57 \end{bmatrix}$$

EXAMPLE

$$A = \begin{bmatrix} 3 & 6 \\ 2 & 4 \end{bmatrix} \qquad B = \begin{bmatrix} 5 \\ 9 \end{bmatrix}$$

$$AB_{11} = \sum_{k=1}^{2} a_{1k} b_{k1} = a_{11} b_{11} + a_{12} b_{21} = 3 \times 5 + 6 \times 9 = 69$$

$$AB_{21} = \sum_{k=1}^{2} a_{2k} b_{k1} = a_{21} b_{11} + a_{22} b_{21} = 2 \times 5 + 4 \times 9 = 46$$

So

$$AB = \begin{bmatrix} 69 \\ 46 \end{bmatrix}$$

# APPENDIX 3: DETERMINANT PROPERTIES

## SCALING A DETERMINANT OR DETERMINANT ELEMENTS

### SCALING DETERMINANT ELEMENTS

Expanding the determinant

$$|A| = \begin{vmatrix} a_1 & b_1 & c_1 & d_1 & \cdots \\ a_2 & b_2 & c_2 & d_2 & \cdots \\ a_3 & b_3 & c_3 & d_3 & \cdots \\ a_4 & b_4 & c_4 & d_4 & \cdots \\ \cdot & \cdot & \cdot & \cdot & \cdots \\ \cdot & \cdot & \cdot & \cdot & \cdots \\ \cdot & \cdot & \cdot & \cdot & \cdots \end{vmatrix}$$

by cofactors found by 'crossing out' the first row and column of elements,

$$|\mathbf{A}| = a_1 \begin{vmatrix} b_2 & c_2 & d_2 & \cdots \\ b_3 & c_3 & d_3 & \cdots \\ b_4 & c_4 & d_4 & \cdots \\ \cdot & \cdot & \cdot & \cdots \\ \cdot & \cdot & \cdot & \cdots \\ \cdot & \cdot & \cdot & \cdots \end{vmatrix} + b_1 \begin{vmatrix} a_2 & c_2 & d_2 & \cdots \\ a_3 & c_3 & d_3 & \cdots \\ a_4 & c_4 & d_4 & \cdots \\ \cdot & \cdot & \cdot & \cdots \\ \cdot & \cdot & \cdot & \cdots \\ \cdot & \cdot & \cdot & \cdots \end{vmatrix}$$

$$- c_1 \begin{vmatrix} a_2 & b_2 & d_2 & \cdots \\ a_3 & b_3 & d_3 & \cdots \\ a_4 & b_4 & d_4 & \cdots \\ \cdot & \cdot & \cdot & \cdots \\ \cdot & \cdot & \cdot & \cdots \\ \cdot & \cdot & \cdot & \cdots \end{vmatrix} + d_1 \begin{vmatrix} a_2 & b_2 & c_2 & \cdots \\ a_3 & b_3 & c_3 & \cdots \\ a_4 & b_4 & c_4 & \cdots \\ \cdot & \cdot & \cdot & \cdots \\ \cdot & \cdot & \cdot & \cdots \\ \cdot & \cdot & \cdot & \cdots \end{vmatrix}$$

Multiplying the first row by a scale factor $\lambda$ increases the value of $|\mathbf{A}|$ by the factor $\lambda$. But each of the cofactors can be similarly expanded in terms of $b_2$ $c_2$ $d_2$ ..., $a_2$ $c_2$ $d_2$..., $a_2$ $b_2$ $d_2$ ..., $a_2$ $b_2$ $c_2$ ..., so that multiplying the second row by $\lambda$ introduces another factor of $\lambda$, and the value of $|\mathbf{A}|$ is increased by the factor $\lambda^2$.

Hence if n rows are multiplied by $\lambda$ $|\mathbf{A}|$ is scaled by $\lambda^n$.

## SCALING A DETERMINANT

Conversely, if a determinant of order n is scaled by the factor $\lambda^n$, its scaled value is equal to the value of the unscaled determinant with every element scaled by $\lambda$.

## THE SIGNS OF THE TERMS OF AN EXPANDED DETERMINANT

Consider the determinant

$$|A| = \begin{vmatrix} a_{11} & a_{12} & a_{13} \\ a_{21} & a_{22} & a_{23} \\ a_{31} & a_{32} & a_{33} \end{vmatrix}$$

Expanding,

$|A| = a_{11} a_{22} a_{33} - a_{11} a_{23} a_{32} + a_{12} a_{23} a_{31}$
$\qquad - a_{12} a_{21} a_{33} + a_{13} a_{21} a_{32} - a_{13} a_{22} a_{31}$

observe that the 1st (row) suffices in each term are in natural order, but the second (column) suffices are some permutation of 1, 2 and 3. The sign of the associated term is given by the number of transpositions of pairs of suffices in the order of the 2nd suffices that are required to obtain the natural order 1,2,3.

For example in,

$$a_{11} \, a_{23} \, a_{32}$$

132 requires one transposition to 123 and

$$a_{13} \, a_{22} \, a_{31}$$

also requires one transposition. These carry a negative sign.

But, for example, in

$$a_{13}\,a_{21}\,a_{32}$$

312 requires two transpositions (231 to 132 to 123), and this term carries a positive sign.

Now consider, say, the 6th term of

$$A = \begin{vmatrix} a_{11} & a_{12} & a_{13} & a_{14} \\ a_{21} & a_{22} & a_{23} & a_{24} \\ a_{31} & a_{32} & a_{33} & a_{34} \\ a_{41} & a_{42} & a_{43} & a_{44} \end{vmatrix}$$

which is given by

$$\begin{vmatrix} \cdot & \cdot & \cdot & a_{24} \\ a_{11} & \cdot & \cdot & \cdot \\ \cdot & \cdot & a_{33} & \cdot \\ \cdot & a_{42} & \cdot & \cdot \end{vmatrix}$$

$$= a_{11} a_{24} a_{33} a_{42}$$

The permutation of the column suffices is 1432, which requires one transposition to the natural order 1234, and the associated term carries a negative sign. The term

$$a_{11} a_{22} a_{33} a_{44}$$

has column suffices already in the natural order. Apparently, then, terms with an odd permutation of column suffices carry a negative sign, those with an even permutation carry a positive sign.

## THE EFFECT OF INTERCHANGING TWO COLUMNS

Each column suffix in every term in the expansion of the determinant receives one transposition. Hence, an odd permutation becomes an even permutation and an even permutation becomes an odd permutation by this operation. Therefore, interchanging two rows of a determinant reverses the sign of the determinant.

## THE EFFECT OF INTERCHANGING TWO ROWS OF A DETERMINANT

Suppose the rows and columns of

$$|A| = \begin{vmatrix} a_{11} & a_{12} & a_{13} \\ a_{21} & a_{22} & a_{23} \\ a_{31} & a_{32} & a_{33} \end{vmatrix}$$

are interchanged, giving

$$|A|' = \begin{vmatrix} a_{11} & a_{21} & a_{31} \\ a_{12} & a_{22} & a_{32} \\ a_{13} & a_{23} & a_{33} \end{vmatrix}$$

Since $|A|$ and $|A|'$ can be expanded by the first column, we have

$$|A|' = a_{13} a_{21} a_{32} - a_{13} a_{22} a_{31} - a_{12} a_{21} a_{33}$$
$$+ a_{12} a_{23} a_{31} + a_{11} a_{22} a_{33} - a_{11} a_{23} a_{32}$$

$$= a_{11} a_{22} a_{33} - a_{11} a_{23} a_{32} + a_{12} a_{23} a_{31}$$
$$- a_{13} a_{22} a_{31} + a_{13} a_{21} a_{32} - a_{12} a_{21} a_{33}$$

$$= |A|$$

Now the same rule that odd permutations of suffices are associated with negative terms and even permutations of suffices are associated with positive terms is here transferred to row suffices, which are now the 2nd suffices. Hence the effect of interchanging two rows of a determinant is the same as interchanging two columns, i.e., interchanging two rows of a determinant reverses the sign of the determinant.

## TWO IDENTICAL ROWS OR TWO IDENTICAL COLUMNS IN A DETERMINANT

Since interchanging two identical rows or two identical columns of a determinant $|A|$ leaves the determinant unchanged yet from above the determinant's sign is reversed, we must have $|A|=|A|'=0$.

## THE SIGN OF COFACTORS
The sign of the cofactor of element $a_{ij}$ in a determinant can be obtained by interchanging row i with row i-1, then with row i-2, and so on until the original row i becomes row 1, involving i-1 interchanges.

Similarly, the original column j becomes column 1 after j-1 interchanges. The first operation produces i-1 reversals of sign of the determinant, the second j-1 reversals. Each interchange multiplies the determinant by -1. Hence, the effect of i-1+j-1 interchanges is to multiply $|A|$ by $(-1)^{i-1+j-1} = (-1)^{i+j-2}$, and so the sign of the cofactor of $a_{ij}$, $|A_{ij}|$ is $(-1)^{i+j-2}$.

ADDING A MUTLIPLE OF ANY ROW OR COLUMN OF A DETERMINANT TO ANOTHER ROW OR COLUMN

Consider the determinant

$$\begin{vmatrix} a_1 & b_1 & c_1 & \cdots \\ a_2 & b_2 & c_2 & \cdots \\ a_3 & b_3 & c_3 & \cdots \\ \vdots & \vdots & \vdots & \vdots \\ a_k & b_k & c_k & \cdots \\ \vdots & \vdots & \vdots & \vdots \\ a_n & b_n & c_n & \cdots \end{vmatrix}$$

Adding a multiple $\lambda$ of row k to the first row:

$$\begin{vmatrix} a_1 + \lambda a_k & b_1 + \lambda b_k & c_1 + \lambda c_k & \cdots \\ a_2 & b_2 & c_2 & \cdots \\ a_3 & b_3 & c_3 & \cdots \\ \vdots & \vdots & \vdots & \vdots \\ a_k & b_k & c_k & \cdots \\ \vdots & \vdots & \vdots & \vdots \\ a_n & b_n & c_n & \cdots \end{vmatrix}$$

$$= \begin{vmatrix} a_1 & b_1 & c_1 & \cdots \\ a_2 & b_2 & c_2 & \cdots \\ a_3 & b_3 & c_3 & \cdots \\ \vdots & \vdots & \vdots & \vdots \\ a_k & b_k & c_k & \cdots \\ \vdots & \vdots & \vdots & \vdots \\ a_n & b_n & c_n & \cdots \end{vmatrix}$$

$$+ \lambda \begin{vmatrix} a_k & b_k & c_k & \cdots \\ a_2 & b_2 & c_2 & \cdots \\ a_3 & b_3 & c_3 & \cdots \\ \vdots & \vdots & \vdots & \vdots \\ a_k & b_k & c_k & \cdots \\ \vdots & \vdots & \vdots & \vdots \\ a_n & b_n & c_n & \cdots \end{vmatrix}$$

$$= \begin{vmatrix} a_1 & b_1 & c_1 & \cdots \\ a_2 & b_2 & c_2 & \cdots \\ a_3 & b_3 & c_3 & \cdots \\ \vdots & \vdots & \vdots & \vdots \\ a_k & b_k & c_k & \cdots \\ \vdots & \vdots & \vdots & \vdots \\ a_n & b_n & c_n & \cdots \end{vmatrix} = |\mathbf{A}|$$

since this last determinant has two rows identical.

The same result is obtained if a multiple of a column h is added to the first column. Since a determinant can be expanded by any row or column, adding to any row or column a multiple of any row or column to a determinant leaves the value of the determinant unchanged

## EXPANSION OF A DETERMINANT BY 'ALIEN' COFACTORS

Suppose the ith row of a determinant $|\mathbf{A}|$ is substituted by a new row $b_{ij}$. Then the value of the new determinant is:

$b_{i1} |A_{i1}| + b_{i2} |A_{i2}| + ... + b_{in} |A_{in}|$.

Now let the $b_{ij}$ be any row of $|\mathbf{A}|$ except the ith row. Then the above expression is the expansion of a determinant with two equal rows, which is zero from above.

Similarly, for a new column, the new determinant is:

$b_{1j} |A_{1j}| + b_{2j} |A_{2j}| + ... + b_{nj} |A_{nj}|$

and if the $b_{ij}$ are elements of any column except the jth, we have an expression for the expansion of a determinant with two columns equal which again is zero from above. Hence, expansions by cofactors of elements in a different row or column, known as alien cofactors, are zero.

## ABOUT THE AUTHOR

Initially educated at a secondary modern school, leaving with 3 'A' levels Peta Trigger went on to university to gain honours degrees in Mathematics, Education and Engineering Science and a post graduate degree in Educational Psychology. Peta continued her education at London University's Institute of Education where she eventually completed her Ph.D and Ed.D degrees in Mathematics and Engineering Science Education. Her research for the Ed.D involved the development of computer programs in teaching undergraduate engineering science.

She is 64 and has lived in Northampton for 17 years.